TO MIKE
IF YOU CAN DREAM IT
YOU CAN DO IT

ROCKETMAN

My Rocket-Propelled Life
and High-Octane Creations

Ky Michaelson

MOTORBOOKS

First published in 2007 by Motorbooks, an imprint of MBI Publishing Company LLC, Galtier Plaza, Suite 200, 380 Jackson Street, St. Paul, MN 55101 USA

Copyright © 2007 by Ky Michaelson

All rights reserved. With the exception of quoting brief passages for the purposes of review, no part of this publication may be reproduced without prior written permission from the Publisher.

The information in this book is true and complete to the best of our knowledge. All recommendations are made without any guarantee on the part of the author or Publisher, who also disclaim any liability incurred in connection with the use of this data or specific details.

We recognize, further, that some words, model names, and designations mentioned herein are the property of the trademark holder. We use them for identification purposes only. This is not an official publication.

Motorbooks titles are also available at discounts in bulk quantity for industrial or sales-promotional use. For details write to Special Sales Manager at MBI Publishing Company, Galtier Plaza, Suite 200, 380 Jackson Street, St. Paul, MN 55101 USA.

To find out more about our books, join us online at www.motorbooks.com.

Editors: Lindsay Hitch and Kris Palmer
Designer: Kou Lor
Cover design: Tom Heffron

Printed in China

Library of Congress Cataloging-in-Publication Data

Michaelson, Ky.
  Rocketman : my rocket-propelled life and high-octane creations / Ky Michaelson.
     p. cm.
  ISBN-13: 978-0-7603-3143-9 (hardbound w/ jacket)
 1. Michaelson, Ky. 2. Rocketry--United States--Biography. I. Title. II. Title: Rocket man.
TL781.85.M53M53 2007
621.43'56092--dc22
 [B]
                                                                                            2007018071

On the cover: Ky models his rocket-powered chair.

On the back cover: Top: The *Pollution Packer* team celebrates after setting 13 state, national, and international records. **Bottom:** Ky stands with the Space Shot 2000 rocket.
All photographs belong to Ky Michaelson's collection.

Contents

Acknowledgments..................................... 5

Introduction 6

Prologue 10

PART I Rockets on Land and Water 14

Chapter 1: Legendary Pioneers of Rocket Car Racing 15

Chapter 2: *Sonic Challenger* 21

Chapter 3: *Pollution Packer* 27

Chapter 4: Bonneville Speed Runs 32

Chapter 5: NHRA Sanctions Rocket Cars 37

Chapter 6: Rocket Racers 42

Chapter 7: John Paxson 45

Chapter 8: Paula Murphy 49

Chapter 9: Ed Ballinger and the *Conklin Comet* 53

Chapter 10: Captain Roller Ball 56

Chapter 11: Space Age Racing Inc. 63

Chapter 12: Side-by-Side Go Karts 65

Chapter 13: Lee Taylor 74

Chapter 14: Human Fly 78

Chapter 15: Kitty O'Neil 83

Chapter 16:	Extreme Rocket-Powered Vehicles	98
Chapter 17:	Legends of Their Time	106
Chapter 18:	Rocketman's Toys	120
PART II	Space Shot on a Shoestring: The CSXT Go Fast Rocket Becomes the First Civilian Rocket to Reach Space	148
Chapter 19:	Space Shot 2000	149
Chapter 20:	Rocket Withdrawal	163
Chapter 21:	The Day the World Stopped Turning	165
Chapter 22:	Mother Michaelson	168
Chapter 23:	Mother Nature's Wrath	171
Chapter 24:	*Primera* Space Shot 2002	178
Chapter 25:	The Journey Continues	192
Chapter 26:	Three Days to Countdown	208
Chapter 27:	Rend the Sky	212
Epilogue:	By the Numbers	225
Appendix A:	Rocket-Powered Vehicles Built by Ky Michaelson	227
Appendix B:	How Hydrogen Peroxide Rocket Engines Work	231
Appendix C:	Official NHRA Rocket Car Rules and Regulations	236

Acknowledgments

This is my first attempt at writing a book, and it has been a real learning experience for me. As a child, I was diagnosed as dyslexic and was far from ever being an "A" student, even though I wanted to be. I started to type this masterpiece at about three mind-blowing words per minute, and it's taken me about two and one-half years to complete. Because I spell every word phonetically, I found myself constantly arguing with spell check since any word spelled differently than it sounds was likely wrong to begin with. Thankfully, I was lucky enough to have a couple of good friends who not only encouraged me to finish this book but also came to my rescue to help me sort out this whole confusing mess.

I would like to thank Nita Icaza for spending a great deal of her time deciphering and straightening out my vocabulary without losing too much of my true character. She embraced the challenge and has hung in here with me this whole time. I would also like to thank Bruce Lee (NOT the motion-picture star you may be thinking of—my friend couldn't karate chop a piece of warm butter in half without hurting his lily-white hands), my fellow racer and "rocket buddy," who commented to me after every rocket car story I told that I should really write a book, and so here it is. Finally, and most important, I would like to thank my mother for bringing me back to this Earth and putting up with a young boy who liked to play with his chemistry set more than do his homework. When the bangs and booms got louder, she would simply say, "Don't get in trouble with the law, and please don't blow your fingers off." She never discouraged me, and her love and support enabled me to take my overwhelming curiosity and interests to heights never imaginable. And best for last, my beautiful wife, Jodi, who inspires me and conspires with me—she's a work of art.

Introduction

This is a very informal book based upon my personal recollection of an evolution that took the racing world by storm and raised the bar for Top Fuel dragsters, the reigning kings of break-neck speeds. Think back to the dawning of the space age, to man's quest for walking on the moon, experiencing g-forces never before dreamed of, and racing cars at speeds never imagined on drag strips throughout the country, and you've now entered the world of hydrogen peroxide rocket car racing. These cars turned into landlocked missiles and gave new meaning to the term "land speed record." When you think of speeds of over 300 miles per hour, you immediately

Ky Michaelson races against his *S.S. Flusher* rocket-powered toilet.

think of the Bonneville Salt Flats. These rocket cars brought their speeds to drag strips in just about every state in the United States. I was fortunate enough to work side by side with those modern-day gladiators, who pushed their own human limits as they burst onto the scene in their wondrous, missile-like vehicles. It was my era, and I dove into it with all my heart, soul, and sweat, hoping, along with everyone else, to achieve goals and records never thought possible. I was witness to the thousands upon thousands of spectators all over the globe who became thrilled, amazed, and educated as they embraced the new sport of rocket car racing.

Although this phenomenon only lasted about 10 years, most track and land speed records achieved during that era still stand today, some 30 years later. It was a time of spills, chills, victories, defeats, and oftentimes tragedy. It is a part of American racing history I feel the need to share, in my own words . . . as I experienced it and remember it.

Since my very early childhood, I've always felt a great need for speed, which is why I believe the good Lord blessed me with my life in this era versus, say, the horse-and-buggy days, although I'm half certain I could have concocted some way to make those old buggies go just a bit faster.

I feel extremely honored and privileged to have played a role in the history of hydrogen peroxide rocket car racing and, even more so, to still be alive

INTRODUCTION

today to talk about it. As I think back to those early days, I'm reminded of the very turbulent times and of the good ones, too. I hope the journey I took as a young man is one that many more will also have a chance to experience and that this time in history will never be forgotten, nor will those brave pioneers, who too often gave their lives to the sport in their quest for victory and the ever-present need for speed for which so many of us seem to live.

It wasn't until I began to write this book that it really occurred to me how many years have passed since we were the kings of speed and just how many of my friends and colleagues are no longer with us. I sought input from some and had the great pleasure of reliving some pretty wild times through our conversations, but in my search for input, I also experienced some sadness in the fact that many of my old racing buddies have aged and their memories just aren't what they used to be. You'll find many stories in here based on my memory and others'. Part of what makes me somewhat eccentric is that I am a pack rat. I catalogued and critiqued my entire racing career and have all the documentation and hundreds upon hundreds of photos to clearly separate fact from fiction. This book is intended to enlighten those who aren't afraid to dream and who share an interest in facing challenges head-on, accepting the outcome, and realizing that nothing is impossible if you set your mind to it and believe.

—Ky Michaelson,
The Rocketman

Prologue

By Ed Ballinger

When Ky approached me and told me he was going to do this book, he asked me if I'd be interested in contributing. I wasn't sure what to say or how I could help out, so he asked me to very simply describe what it's like to drive a rocket car. He thought it would help readers to appreciate the sport more fully if they could hear it described by someone who was there, who'd taken the ride, time and again, and survived. And so . . . I'm very humbled and privileged to offer to you the experience of a lifetime, in my own words.

What's It Like to Drive a Rocket Car?

Like most other things in life, driving a rocket car has its good days and its bad ones. The bad days are the safest, because on those days you're truly scared, and when you're scared, everything, no matter how trivial, gets your full attention. The speed or the elapsed time (ET) of the rocket car is not the least bit affected, but you are.

The good days are those macho times that seemingly have no limits whatsoever. You load the fuel right to the very top of the fuel tank just to make sure there's enough to warm the rocket engine so it will operate at its maximum power right through the timing lights. I found out after a couple of close calls that it's not good to be macho. As the old saying goes, "There are old men, but there are no old bold men."

PROLOGUE

Ed Ballinger in the *Conklin Comet* with the chute out.

    Race day arrived. Before I'd make my run, I'd always spend a little time by myself relaxing and making sure my head was in the right frame of mind. When it was time to put on the fire suit and get seated in the cockpit of the rocket car, it was time to get serious. One of the crew members would help with the six-point seatbelt, but we'd go one step further and take a vice grip, clamp it to the end of the seatbelt, and pull as hard as we could because the chutes would hit so hard, if the belts weren't tight enough, they'd leave black-and-blue marks on my shoulders. Our communication system was simultaneously broadcast over the track P.A. system so everyone could hear our conversation as we went through our checklist together. This not only added more drama to our show but also educated the spectators on how the rocket system was prepared before I was pushed to the starting line.

    Once the car was aimed down the track, Ky or one of the crew members would instruct me to open up the main air-actuated safety throttle valve. Once I did that, I would start to slowly pressurize the hydrogen-peroxide fuel tank with nitrogen by turning the dome-loader regulator control. When the tank pressure reached 300 psi, Ky would give the signal to push down slowly on the throttle valve and let a little hydrogen peroxide into the rocket chamber. As soon as the peroxide hit the silver catalyst pack, the motor would begin to snap and pop. White, superheated steam would begin to flow out the rocket nozzle. I'd wait for Ky's verbal command to, "hit the throttle," a couple of times until the rocket reached its full operating temperature. At this point, a thousand things would

> **The spectators would give a NASA-type countdown, starting at 10, and when they'd reach 1 . . . it was showtime!**

be going through my mind as I looked down the track. Things like, "Will I have a parachute failure? Will the tires be able to be stressed beyond their manufactured capabilities one more time?" And finally, the worst thing of all: "Will this be my last run in the car?" It was pretty hard not to let those negative thoughts go through my mind when the fact was we'd lost a number of rocket car drivers in a very short time. I always just wanted to be sure I wasn't included in that list. When the temperature in the motor reached about 1,300 degrees, the car was then pushed to the starting line.

I was now mentally ready to go. The spectators would give a NASA-type countdown, starting at 10, and when they'd reach 1 . . . it was showtime! I'd take one big breath and smash the throttle down as hard as I could.

Ky Michaelson pushes Ed Ballinger up to the starting line.

PROLOGUE

Now the fun began. The acceleration was indescribable—like going down the biggest roller coaster I'd ever been on, times 10. Because of the g-forces, it felt like the steering wheel was going to be ripped right out of my hands. In the first second, I accelerated to over 100 miles per hour!! My vision blurred, the acceleration became even more intense, and when I'd pass a light or telephone pole, it seemed to bend as I went by. Everything in front of me came at me in a jerking motion. The roar of the rocket engine was constant; the only change was the intensity of the body banging against the chassis and the rising howl of the wind that relentlessly increased in its shrieking tempo. At about 200 feet before the timing lights, I'd shut the rocket off and pull the primary chute at the lights. The reverse force from the chute is about 14 gs and the feeling is incredible. I would describe the run as an "eyeballs in and eyeballs out" experience. It almost felt like the car was driving down a tunnel right into the ground. The overall rush always led me to say, "I cannot wait to do this again."

I remember one time when I drove Ky's *Mattel Rocket Car* in September 1978. He decided to sell the car and concentrate on building a land-speed-record car instead. The Mattel car had a 7,500-pound-thrust engine in it and was without question the most powerful rocket car I had ever driven. Ky told me to shut the car off 300 feet before the lights and pull the main chute. The chute at that speed opened really fast, and I hardly had time to take hold of the wheel before it snapped my head down. I did exactly what he said, and the g-forces were phenomenal. I managed to stop the car just 25 feet from going off the end of the track and was clocked at 347 miles per hour through the lights with the primary chute out for 200 feet. I'm sure the car was going over 400 miles per hour before the chutes fully deployed.

Ky and the crew drove right down the strip chasing me, and Ky was all out of breath and laughing. The first thing he said was, "Ed, did you feel your NHRA rocket license explode in your wallet as you passed through the finish line?" At the time, though, I wasn't the slightest bit concerned that I'd more than likely lost my license and my driving privilege. All I cared about was that I had just set a new quarter-mile speed record of 347 miles per hour and was now the new king of quarter-mile speed! More amazing than that, I'd done so safely and walked away without a scratch.

Part I
Rockets on Land and Water

Chapter 1

Legendary Pioneers of Rocket Car Racing

The hydrogen-peroxide rocket car story actually began back in the mid-1960s, when Ray Dausman, who worked as a research technician, attended a national drag race. His sharp, determined mind set to spinning as he figured there had to be a better way of achieving the ultimate quarter-mile speed. Ray had been hugely interested in rockets as a hobby when he was just a young boy and was a member of the American Rocket Society. He'd built a number of experimental rocket motors in the basement of his house, and while most mothers wouldn't have allowed this type of activity, Ray was blessed to have the same type of mother I had, who was very supportive of his willingness to learn through experimentation.

The first-ever, one-of-a-kind hydrogen-peroxide-powered dragster.

ROCKETMAN

While attending Purdue University, Ray designed and tested two liquid-fuel and two solid-fuel rocket engines. It wasn't long before young Ray's daydream became reality as he and Dick Keller, his trusted colleague from ITT Research Institute in Chicago, set out to achieve his vision. In hopes they could upscale a drag racing car guaranteed to leave the high-speed piston pushers in the dust, they set out to build a small 25-pound-thrust prototype hydrogen-peroxide rocket motor. While they worked on the motor, they expanded their vision by putting a three-step program into place:

The *X-1* rocket car chassis and engine.

1. Clearly demonstrate their ability to design, build, and successfully test-fire their prototype.
2. Design and build something large enough to power a dragster.
3. Go for the ultimate creation . . . use rocket power to beat the existing land-speed record held by famous race car driver, Craig Breedlove, whose car was jet-engine powered.

Those were the plans, and nothing and no one was going to stop them from reaching their goals. Keller and Dausman successfully test-fired their motor in fall 1964, using a bathroom scale to measure the thrust of the rocket. They recorded this history-making event with their home movie camera, which allowed them to demonstrate this feat with ease. The motor performed close to their specifications, but there was a significant glitch. Not all of the hydrogen peroxide decomposed and, because the rocket nozzle was pointed upward, the undecomposed peroxide shot straight up into the air and then dropped in a shower all over Ray's head. Needless to say, he ran like a track star to the sink and doused his head with water, avoiding a potentially serious injury. That scare, and the fact that they were extremely short on cash, did not dissuade them, and they forged ahead with the second step of their plan.

They had the vision, but they knew they needed help to proceed, so they sought the best in the business, professional race car designer and driver, Pete Farnsworth. The three visionaries formed a partnership called BFK Enterprises, which was a very rock-solid, talented group of men that wasted no time moving forward. They decided to start by designing the ultimate rocket car dragster, and what a design it was. The thing was completely state of the art, and once it went from their heads to paper, they were set to start building.

As Farnsworth began building the chassis, Keller and Dausman got busy and enlisted another very talented man by the name of James C. McCormick to assist with the rocket and catalyst pack design. James had worked for 13 years at FMC Corporation, developing hydrogen-peroxide power systems, and now headed up his own consulting firm. With his help, they designed a 2,500-pound-thrust motor, a huge upgrade to their previous prototype. The majority of the machine work was done at the Illinois Institute of Technology (IIT), and the motor was completed in 1965. Ray told me that this was the largest hydrogen-peroxide motor ever built, and I have absolutely no reason to doubt him.

The car itself took quite a bit longer to complete than they had planned. Every weekend they could, Dick and Ray drove up to Milwaukee from Chicago to help Pete work on it. Money was scarce, and Dick and Ray weren't able to contribute as much as they'd have liked to, so the shoestring budget wound up being financed mostly by Pete and his wife, Leah. Minus the streamlined aluminum body, the chassis was completed in July 1966, nearly a full year after they'd started. They had successfully created the first-ever, one-of-a-kind hydrogen-peroxide-powered dragster, and it was truly magnificent. The 750-pound car, dubbed the *X-1 Special*, looked more like a Formula One–type race car than a dragster, with its fully enclosed driver's compartment and the latest and greatest technology available. It had front and rear independent suspension with four-wheel disc brakes and three parachutes for additional stopping power. Its most unique attribute, though, was that it was powered by 90 percent hydrogen peroxide, and its mighty little rocket engine produced 2,500 pounds of thrust.

With the funds now fully depleted, the time had come to find a sponsor, so they turned to the Rislone Company, which graciously agreed to team up. Although it was not a monetary relationship, Rislone used the team in its ads, and the company sent the team and the car out to California for a press conference, which gave them an incredible amount of exposure. Fees from drag strip bookings plus money from their day jobs kept the project alive.

With an official name and an official sponsor, the time came to see what this fine machine could do, and Dick stepped up to the challenge. He was the first to drive the awesome little *X-1*. He fired the engine for only two seconds on that first hair-raising run. That was all it took for him to realize he needed to

The *Blue Flame* land-speed record car.

find a professional driver, so Pete referred him to an old friend, Chuck Suba, of Calumet City, Illinois. Suba was very seasoned and experienced at driving dragsters and jet-powered cars and eagerly agreed to drive, but only under one condition: that he be first in line to drive their proposed land-speed record car if they made it to phase three of their plan and ever got it built.

On Chuck's first run, he drove down the quarter-mile at a speed of 203.39 miles per hour, which was absolutely awesome given the fact that the car wasn't even completed yet. Pete finished the body in April 1967, and the team continued to test the car the rest of that summer. When Labor Day weekend rolled around, it was no longer just a test. They showed up at a drag strip and challenged Walt Arfons' *Green Monster* jet car to a side-by-side race. Walt agreed to the race under one condition: that they give him a half-second head start. They agreed, so the officials were notified. The crowd went nuts as the two cars lined up side by side. Even with the half-second advantage, Suba shot past the *Green Monster* to clock an elapsed time of 6.32 seconds for the quarter-mile race. That was an incredible start to what became a hugely successful team effort. The *X-1* made exhibition runs in Wisconsin, Illinois, Indiana, and Oklahoma for about two years. According to Pete Farnsworth, the car's highest recorded quarter-mile speed was 265.480 miles per hour, and the lowest ET was 5.90, achieved in Oklahoma City in 1968. That was also the last time the *X-1* was ever run on a drag strip. The team was ready to move on to phase three of the plan: it was time to build a land-speed record car.

Ray Dausman holds a small 25-pound-thrust motor next to the *Blue Flame*'s 22,000-pound-thrust motor.

Once again, they came up with a design like none other. The back end of the vehicle resembled a jet fighter, complete with a fully enclosed driver's compartment similar to an airplane cockpit, attached to a cylinder-shaped body that looked like a missile or rocket on its side. It was during this time that they also changed the company's name to Reaction Dynamics Company, with Pete as president, Ray as vice president, and Dick as secretary/treasurer. On this go-round, their own company built the chassis.

They contracted with Galaxy Manufacturing of Buffalo, New York, to build the rocket motor. As their new creation neared completion, they once again sought sponsorship. They found a major sponsor when The American Gas Association (AGA) and about 30 of its member companies came aboard and agreed to provide the project with financial support in the form of advertising dollars (not research funding). The gimmick that truly attracted the AGA was the use of liquefied natural gas (LNG). This vehicle was designed to produce 22,000 pounds of thrust. The rocket motor ran on 90 percent hydrogen peroxide, and the LNG was used as fuel to burn with the oxygen released from the peroxide decomposition. When the motor was initially tested in July 1970 at Great Lakes Dragaway in Union Grove, Wisconsin, it produced 16,500 pounds of thrust. The project was originally estimated to cost about $70,000 but ended up costing over $500,000. Even so, the *X-1* soon became the predecessor to the new land-speed record car, appropriately named the *Blue Flame*, which was a great symbol for all its sponsorship companies.

Very sadly and very tragically, Chuck Suba was killed in an accident in a conventional dragster before he ever got a chance to drive the new car, and it was a devastating blow to the team. With heartfelt sadness, they knew they had no choice but to forge ahead without him. They interviewed and screened a dozen professional drivers to take the helm of the *Blue Flame*, including Don Garlits. They finally chose Gary Gabelich, a well-known race car driver from the west. Gary had some experience in aerospace, which they thought would be very helpful; in fact, at the time, he was actually testing a space suit for Mercury astronaut Wally Schirra because he and Wally were basically the same size. During this transition, the team unfortunately experienced personal conflict, and Ray Dausman dropped out before the car actually ran.

Gary Gabelich went on to set a new land-speed record of 622.407 miles per hour when he drove the *Blue Flame* on October 23, 1970. They were the only team to achieve the land-speed record in only one season on the Bonneville Salt Flats. They had done what they had set out to do: break Craig Breedlove's *Spirit of America* record of 600.601 miles per hour. They couldn't have known at the time that they'd hold that record for 13 years, which was another record in itself. The *Blue Flame* is now a museum piece, a gallant symbol of rocket-car history to be cherished forever.

Gary Gabelich with a wooden model of the *Blue Flame*.

Dick Keller holds the wind-tunnel model of the *Blue Flame*.

Chapter 2
Sonic Challenger

Ky Michaelson gets ready to test the *Sonic Challenger* configured with snow tracks on.

In 1969, I happened to read an article on the *X-1* dragster in one of my car magazines, and I immediately recognized the name Pete Farnsworth. Because the article talked about his latest *Blue Flame* project, and I had also evolved into the wonderful world of rockets, I decided to give him a call. I figured we'd have a lot to talk about, and I was right. During our conversation, he asked me if I'd be interested in buying the old *X-1* because they were experiencing some financial difficulties with the project. Well, at the time, it wasn't something I could afford on my own, but seeing as how I was employed as a project engineer for the Couparral Company, manufacturer of Sno-Pony snowmobiles and other recreational vehicles, I came up with a plan. I talked Tony Fox, the owner of the company, into buying it as a company investment and as a great opportunity for lots of publicity.

In fall 1969, we flew out to Milwaukee for a close-up look at the *Blue Flame* and then also took possession of the *X-1*. We had it trucked back to

Minnesota, where it immediately took on a whole new persona. I personally put it through the transformation process, modifying it from a rocket car to a rocket snowmobile . . . definitely the first of its kind. I removed the front wheels and replaced them with skis, replaced the rear wheels with snowmobile tracks, and painted it Sno-Pony yellow. When it was all spit-polished, I proudly dubbed it the *Sonic Challenger*.

By that winter, we were ready to try it out. In February 1969, there was a snowmobile race at the Brainerd International Speedway in northern Minnesota, so we packed up and headed north.

When we arrived, it was a crisp 20 degrees below zero, which immediately presented us with our first challenge. We had to keep the barrels of hydrogen peroxide from freezing, so we used tank heaters. We weren't going to let the frigid temps get us down, though, so we forged ahead with great optimism. However, after a couple test runs, we soon realized we had some significant problems. Right off the bat, we had to remove the tracks, because on every run I made, as soon as I shut off the rocket motor, they had so much drag they'd cause me to spin out, which wasn't good at all. It was a lot more nerve-wracking an experience than I cared to endure. Although it was a very educational and exciting day at the track, and we managed to make a lot of very curious fans happy along the way, we returned to the Sno-Pony plant and replaced the tracks with skis, thinking we'd solved the problem. Sadly, the next test run proved us wrong again. We had ultimately replaced one significant problem with yet another, because now I had no brakes and had to rely solely on the parachutes for stopping.

I was hooked on this newest creation, but sometimes raw enthusiasm just isn't enough. Before we even had a chance to run it again, the Couparral Company was bought out by the Studebaker-Worthington Corporation, which immediately changed the business name to Sports Power, Inc., and also took immediate possession of our awesome sled. While the deck certainly seemed stacked against us, I soon did my homework and learned that the corporation also happened to own STP and was in fact very heavily into racing. Though not involved in snowmobile racing, the company sponsored the Andy Granatelli Indy cars and Richard Petty's NASCAR racing team. I felt I could spark their interest and try to expand their horizons.

I set up a meeting with the upper management and convinced them that we could get a lot of publicity in the snowmobile industry if we could set a world snowmobile speed record with the *Sonic Challenger*. Getting their support and sponsorship was actually pretty easy; convincing the United States Snowmobile Association to let us go after the world record with a rocket-powered snowmobile proved to be much more of a challenge. After many grueling weeks of discussion, they finally thought it would be good for the entire industry to get some worldwide publicity, so they gave us the go-ahead.

On February 14, 1970, a full year after our initial tests, we traveled all the way up to Lake Champlain near Burlington, Vermont, for a huge event. Our intent, of course, was to set a new record, and even though the temperature was 10 degrees below zero, we were successful and set the record at 114.50 miles per hour. Of course, today that doesn't sound fast at all, but you have to remember: we were the true pioneers of snowmobile racing. Our success that day certainly didn't come without some pain. I tore the rear ski completely off after hitting a large chunk of ice and spinning out, which created a lot of damage to the semi-monocoque chassis, and that was on my first run of the day. I knew the *Sonic Challenger* was capable of going over 250 miles per hour; I just needed another chance at it. While we waited for the world to catch up and for others to follow in our footsteps, I continued racing ordinary snowmobiles, which proved to be a mistake on my part. I ended up breaking my left arm and wrist in a bad accident, which put me out of commission.

In March, we were invited to run the *Sonic Challenger* at the Yellowstone National Park speed run event. While it had taken some time for my wounds to heal, I was more than ready and absolutely thrilled as the promoters painted the picture. I was told that the event was held on an old abandoned airstrip, which sounded ideal. I accepted the invitation on behalf of the company, and the team and I packed up and headed out.

Amazingly, my vision of the event's ideal conditions was quickly erased as we approached the location. What lay in front of us was anything *but* ideal, and my jaw literally dropped. The airstrip they had bragged about was definitely abandoned, all right. It was also buried a good 15 feet under the snow, and that wasn't even the worst of it! The promoters had used a bulldozer to cut a 12-foot path with 4-foot walls for the track. It looked more like a bobsled track than anything else, and as I stood there looking down the tunnel of terror, all I could think of were the spectators. I wondered how in the world they'd possibly be able to see anything, as there were no bleachers, and more important, how they'd stay out of harm's way. I immediately voiced my concerns, and they very confidently assured me that they'd keep the crowd back. With much hesitation, I took their word for it and just hoped for the best.

The morning of the event, we arrived early and were amazed at how big it promised to be. Thousands of people began trickling in, and I really started having misgivings again. I was still very uneasy about the track and just wasn't convinced these guys could truly handle crowd control. My gut instinct proved to be right. As I got into position and began warming up the rocket motor at the starting line, I saw the spectators starting to move closer and closer to the course. The warning bells began ringing in my head, and all I could do was hope they had things under control. That couldn't have been further from the truth, because the second I pushed on the throttle, the

spectators converged on the course. I was going about 180 miles per hour when I shut the motor off because the *Challenger* had begun going up the 4-foot banks, and I was traveling from one bank to another. All I could see was sheer horror on the spectators' faces and fear in their eyes as I deployed the chute to slow down. By the time I hit the timing lights, I was traveling a very disappointing 60 miles per hour, but I was just thankful that no one had been hurt. I had been scheduled to make two runs, but after that incident, I decided it was best for everyone not to run again. We loaded up our semitruck and headed back to Minnesota.

Sadly, because no one else even came close to breaking our original record, the *Challenger* was sent off to the old Studebaker manufacturing plant in South Bend, Indiana, where it sat in cold storage for over two years. That was not to be the end to my disappointment. Because the snowmobile marketplace had become saturated, with over 35 other manufacturers mass-producing them, Studebaker decided to get out of the business altogether. I couldn't have known it at the time, but ironically, that major turn of events was the pivotal turning point in my life. I soon lost interest in racing sleds, which turned out to be my big break, because I had a complete change of heart. I gave it some serious thought and decided I wanted to become a full-time professional drag racer and build an exhibition rocket-powered dragster. I figured I could use the *Challenger*'s chassis, and since I knew it was just taking up space in storage, I managed to talk the company into selling it to me for $2,500.

That weekend, I hooked up my trailer and headed down to South Bend. The place was easy to find, and I got pretty excited about seeing the old sled again and hoped it was still in good shape. I hopped out of the truck and headed toward the plant, where my mood quickly changed. As soon as I walked into the place, I was overwhelmed with a really eerie feeling. I walked down the old assembly line, and it felt just like I was walking through a ghost town. I started to realize what a true shame it was that they were no longer making cars there; where hundreds of gainfully employed people had once stood was now just a barren place full of old memories and nothing more. As I slowly walked around inside the abandoned building, every step an echo, it hit me like a rock that I truly wasn't interested in the *Sonic Challenger* anymore. I felt it had perhaps found its final resting place, and since there was no one around there to help me load it anyway, I decided not to take it. I found it resting in a dark corner and, without hesitation, climbed in and sat for a good, long while. I remembered all the fun and all the things I'd learned while running it, but my spark was just no longer there. I walked away that day with only the rocket motor and the propulsion system and left the rest behind, closing an exciting chapter in my young life.

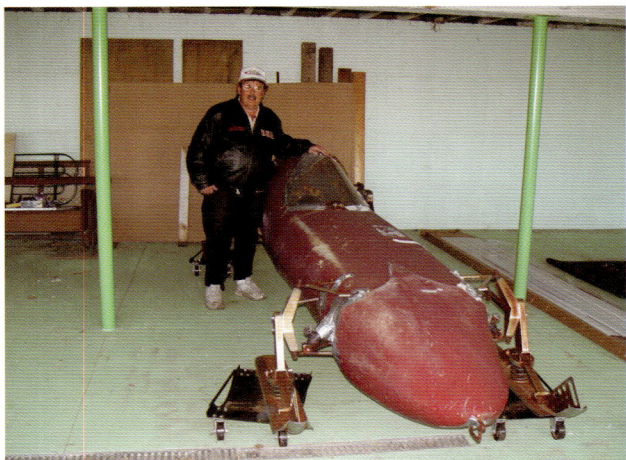

The author with the *Sonic Challenger* after it was rescued from a scrapyard in Michigan.

Thirty years after that very memorable day, I was giving a talk at an antique snowmobile club when one of the members from Michigan raised his hand and said, "I think I know where the *Sonic Challenger is*. I saw it on top of a semi that was used for a sign at a snowmobile scrapyard in Michigan."

I couldn't believe what I'd just heard.

The second I got in the door at home, I immediately started calling around. You can't begin to imagine my excitement when, lo and behold, I actually found it! In my own mind and my heart, I could only think of one thing to do. I felt that the old *Sonic Challenger* should go back to its original owner, the creator and builder of the *X-1*, Pete Farnsworth, so I gave him a call right away. I'm happy to say that at the writing of this book, he is now in the process of restoring this wonderful piece of history. As soon as he got it back, I went to Milwaukee to see it, and it looked to be in pretty bad shape. However, I know that Pete is a true craftsman, and he will restore it to its original perfection. I'm really looking forward to seeing it again one day, the way I remember it.

Chapter 3

Pollution Packer

Ky Michaelson makes the first run in the *Pollution Packer* at Minnesota Dragways.

O ver the course of about 15 years, I raced cars and motorcycles, both as a hobby and semiprofessionally, and also campaigned top gas dragsters around the country. Fortunately for me, I even had the opportunity and privilege to race in the NHRA Division Five Points Race before they decided to drop top gas as a professional class and replace it with Pro Stocks. I wanted to remain competitive in some way and found myself with a rocket motor in one hand and a rear-engined dragster in the other. At the time, it made good sense to me to yank the supercharged 427 out of my dragster and replace it with the rocket motor from the *Sonic Challenger*, so in the winter months of 1972, that's what I did. I made many modifications, including replacing the 12-gallon fuel tank with a 24-gallon tank. From there came the propulsion system: I changed it completely to increase the flow of hydrogen peroxide, which in turn increased the thrust and burn time.

The *X-1* rocket chamber with an all-new fuel system and 24-gallon tank, installed in Ky Michaelson's dragster.

Just about the time I finished my new dragster, Tony Fox, my old employer, who also happened to be my neighbor, stopped by to check it out and asked me what my plans were. I told him I was gonna go after the world quarter-mile record with it, and he immediately became almost as excited as me. Being the businessman he was, he quickly realized that this car had the potential to get a lot of national publicity, so without hesitation, he told me if I came back to work for him as an engineer to help develop his line of trash compactors, his company would sponsor my rocket car. I have to say I was *very* reluctant to consider working with him again because of some past disagreements we had had, but, against my better judgment, I let him convince me.

The first thing that changed was the identity of my car. It was repainted and renamed the *Pollution Packer*, after his trash compactor company. Soon afterward, we became known as the Pollution Packer Rocket Team, and the company went full throttle on its sponsorship, buying a brand-new semi-truck painted with the company logo and the car's colors to transport the dragster around the country. The crew traveled in style, too, with our own personal Bell Jet Ranger Chopper, complete with pilot.

POLLUTION PACKER

Bill Fredericks stands by the *Courage of Australia*. The car was driven by John Paxson at the Great Lakes Dragaway during the Rocket Jet Car Nationals.

At the time we were entering the arena of rocket car racing, Bill Fredericks—the former owner of the Valkyrie jet car and chief engineer for Romatec Laboratories from the West Coast—built a three-wheeled, 27-foot-long rocket car that weighed in at 1,100 pounds in just 11 short weeks. It looked very similar to the *Blue Flame* world land speed record car. He received a sponsorship from Wynn's Oil Company, and they named it the *Courage of Australia*. Vic Wilson drove the car and set an all-time reported speed record of 311.41 miles per hour in 5.107 seconds on his initial test runs at Orange County Raceway in Southern California on November 11, 1971. That record broke the 265.80 miles per hour in 5.90 seconds record set by the *X-1 Rislone* rocket back in 1967. After setting the new record here in the United States, Bill shipped the car over to Australia in January 1972 to make exhibition runs at a number of drag strips there, where he set the Australian record at 243 miles per hour in 6.52 seconds.

The *Pollution Packer* made its first test runs in May 1972 at Minnesota Dragways. I quickly discovered the tremendous amount of power the car really had and also how much easier it was to drive than a car with a reciprocating engine. I was having a ball, so I got a hold of my old school friend and fellow drag racer, Dave Anderson, and invited him to come out and watch us test the car. Dave drove his own Top Fuel dragster for a number of years and was a member of our Gopher State Timing Association (GSTA) hot rod club, so he was definitely game for watching. Well, it wasn't long before he asked me if he could drive, and I was more than happy to let him jump into the driver's seat. His very first run was well over 200 miles per hour, shutting the car off 300 feet before the timing lights. That was it—he was hooked on rocket-powered cars from that day forward. Actually, I think he was getting tired of dodging superchargers and other parts that were constantly flying off his dragster. He asked me if he could have the honor of driving the car at our upcoming appearance in Union Grove, Wisconsin, at the Great Lakes Dragaway Rocket and Jet Car Nationals to be held on Labor Day weekend 1972. Bob Metzler, the track promoter, advertised the event as the big showdown between two rocket cars, the *Pollution Packer* and the *Courage of Australia*, even though we weren't actually scheduled to race each other. The real challenge was to see who would set track speed and elapsed time records.

Dave made his first exhibition run against Tommy Ivo in his Top Fuel dragster. Needless to say, Dave left Tommy in a big cloud of smoke, and the crowd went absolutely crazy. Because of the extreme humidity that night,

POLLUTION PACKER

the exhaust of the *Pollution Packer* engulfed the entire racetrack, and it hung there thick as fog for about 5 minutes or so, which got the crowd even more excited. For a good 15 minutes, they continued jumping up and down, tears running down their faces from the hydrogen peroxide residue left in the cold, humid air. On that first run, we set the new track record for Union Grove at 282.37 miles per hour in 5.68 seconds. Keep in mind this was 1972 and Top Fuel dragsters were running 240 miles per hour. We knew then that we hadn't even tapped into the full potential of the car.

    The next day, we confidently challenged Bill Fredericks to a side-by-side race for a sizeable amount of money. To this day, I'll never know why, but he refused to take his car off the trailer and run against us. Here he had claimed to have the world's fastest quarter-mile car, we were there to prove him wrong, and he wouldn't accept the challenge. I have always had a lot of respect for Bill and his mechanical genius and can only guess that on that particular day, he had exercised better judgment than we had. Looking back, in 1972, no one really knew the full potential of each of those rocket cars. Had we lined up side by side, both loaded for bear and going all out, disaster could have very easily struck, and someone could have been seriously injured or killed. We left the track with the record-breaking 280.37 miles per hour in 5.68 seconds. Not bad for our debut.

Chapter 4

Bonneville Speed Runs

Ky Michaelson and Dave Anderson hold a trophy after record runs.

We headed back to Minneapolis, very pleased with the performance of the car, and decided our next feat would be to go after the world acceleration records under the Federation Internationale de l'Automobile (FIA) sanction and the United States Auto Club (USAC). We were so completely confident in ourselves and in the car's ability that we forged ahead with plans for the monster party we were gonna throw to celebrate our victory. We actually put together a promotional package that was beyond belief and was a sure-fire way to place the Pollution Packer Rocket Team and its car in more limelight than any other race team out there, as well as bring in gads of money for the company. The package was presented to all the *Pollution Packer* distributors, who were told if they bought so many compactors, we would fly them out to the Bonneville Salt Flats, pay for their food, lodging, and transportation, and give every one of them $100 worth of quarters for gambling. In addition to our distributors, we also invited some of the hottest guys in the racing business. Invitations were sent to Dean Moon, Jim Deist, Craig Breedlove, Gary Gabelich, Art Arfons, Doug Rose, Fred Sibley, Roger Gustin, John Paxson, and many more. We also included Hollywood celebrities James Brolin and Dale Robertson. Joe Petrali from the USAC was there to oversee the entire event and to represent the FIA of Paris; Jack Hart of the National Hot Rod Association (NHRA) was there as well. We even flew in a band called the Fiftieth State from Hawaii for the monumental occasion. Our efforts were not in vain, and it was absolutely, without question, the biggest and best party ever thrown on the Flats, and I believe that still holds true today.

The author, Dave Anderson, Paula Murphy, Gary Gabelich, and Art Arfons take a minute between runs.

The salt is so flat that you can see the line of the telephone poles disappear over the horizon because of the curvature of the Earth. Because this area is so open and isolated, we had to bring a ton of tools, spare parts, and equipment with us to support our speed record attempts. As it turned out, while we were fueling the *Pollution Packer* (with 90 percent hydrogen peroxide), one of the crew members accidentally stripped a large stainless-steel fitting in the fuel tank. Wouldn't you know it? We didn't have a 1 3/4-inch pipe tap to repair it. I instantly got a real sinking feeling in my gut as I realized we were out in the middle of the Flats, with hundreds of guests, and a stinking stripped fitting was gonna prevent us from running the car, so I immediately sent three of my crew members in three different directions to find a tap. Salt Lake City was about 125 miles away, and I didn't think Wendover, the small nearby town, would have a hardware store that would carry that big of a tap. I was right; however, the owner of the store recommended we try the nearby salt mine. Bingo! Sure enough, they had one in the repair shed and were kind enough to let us borrow it.

After we repaired the tank, I suddenly realized we had a new serious problem. All four of the tires had lost a considerable amount of air; in fact, one of them was totally flat. Cragar Wheel Company had made us a special

set of spun two-piece aluminum racing wheels to hold the high-speed Goodyear racing tires, so I wasn't quite sure how to make repairs. (The aluminum wheels we had were the predecessors to the Cragar S/S wheels you see on today's cars.) I called the Goodyear racing rep, and he told me their racing tires did leak air, but the seals in the rims had probably also leaked. Since there was really nothing we could do about that probability given our location, we just checked the tire pressure after each run and kept refilling them with nitrogen. Our persistence paid off as one by one we set world

BONNEVILLE SPEED RUNS

acceleration records, which included the quarter-mile at 158.8 miles per hour in 5.666 seconds, the 500-meter at 173.9 miles per hour in 6.31 seconds, and the kilometer at 234.7 miles per hour.

After Dave set the men's records for ET and miles per hour categories, we put the extremely talented Paula Murphy in the car, and she proceeded to set all of the women's records with it as well. Under international rules, you have to make two runs in one hour on the same measured course to set an official record, and while these speeds don't sound fast, they are the average speed of the runs. In all, we set 13 sanctioned national and international speed records before finally returning home to Minneapolis.

Pollution Packer with the chute out after setting a world record.

Paula Murphy helped the *Pollution Packer* team set 13 sanctioned national and international speed records.

ROCKETMAN

Shortly upon our return, we received an interesting telegram from the office of the one and only Howard Hughes, inviting us to bring the *Pollution Packer* to Las Vegas and stay at his new Landmark Hotel. He had apparently heard of our car and the team and wanted an opportunity to show his appreciation for our accomplishments. Upon arrival, we were picked up at the airport in one of his limos. I asked the driver the last time he had seen his boss, and he told us it had been about 15 years. He continued on, stating that Mr. Hughes was a very private man. We were truly in awe to have received the invitation and had a remarkably great time during our stay. The *Pollution Packer* was on display and received a lot of attention, and we were fortunate enough to have dinner with the lovely Ms. Phyllis Diller and Mr. Jimmy Dean. We were entertained every night, either attending a stage show or some other big event. All of our food, lodging, and shows were put on a tab and picked up by Mr. Hughes. He certainly did nothing but demonstrate his great generosity, and our only regret was not having the chance to meet him in person to thank him for all the kindness, but I guess that was just his way.

The *Pollution Packer* team celebrates after setting 13 state, national, and international records.

Chapter 5

NHRA Sanctions Rocket Cars

Dave Anderson and the *Pollution Packer* make history again with a 4.99-second ET and a speed of 322.00 miles per hour.

We began negotiating with Wally Parks and the NHRA to get rocket cars approved to run on NHRA tracks. For many years, they wouldn't allow jet cars to run on their sanctioned tracks. The NHRA believed they were too heavy and, because of the high rpm of the jet engines, much too dangerous. Their concern was that the motor might ingest something through the intake, prompting a catastrophic engine blowup.

ROCKETMAN

Bill Farmer, Tony Fox, Wally Parks, and Gary Rosenfield check out the *Pollution Packer* while Dave Anderson sits in the car.

After several months of arguing our point that rockets had no internal moving parts and were considerably lighter and much easier to stop than the heavy jet cars, they finally gave us the go-ahead. They put Bernie Partridge in charge of the new rocket racing program, and one of his first tasks was to lay down the rules. Luckily, Bernie let us work with him, as well as the consultant the NHRA hired, Andy Kubica, who still managed to set up some very stringent rules for us to go by. They set an absolute speed limit on an NHRA track of 275 miles per hour for Funny Cars and 330 miles per hour for dragsters, and stated that a minimum elapsed time of 4.50 must be observed or your license could be revoked. In addition, no side-by-side rocket car racing would be allowed. The *Pollution Packer* and Bill Fredericks' *Courage of Australia* were the first two cars approved, and John Paxson received the very first NHRA rocket license to drive, number 001. Our drivers, Dave Anderson and Paula Murphy, received the second and third licenses, numbers 002 and 003, respectively.

Dave and the *Pollution Packer* had the honor of being the first rocket driver and car ever to run at an NHRA-sanctioned event. The car made its debut at the Winternationals in 1973 and set the NHRA world record at 297.02 miles per hour in 5.35 seconds in front of tens of thousands of screaming racing enthusiasts. It was the first five-second run on an NHRA track. Right afterward, Paula Murphy jumped in and set the women's mile-per-hour record at 258.62 miles per hour in 6.00 seconds. I was told there were a number of big-name drag racers who were not happy with the rocket car stealing their thunder and getting the majority of the press coverage, including a feature story on the *Wide World of Sports* television show. At the same time, however, there were also rumors floating around the pits that after the rocket car ran, the traction on the track was much better. I never knew if that was true or not, but the *Pollution Packer* did a great job of blowing all the debris off. I'm thinking maybe the hydrogen peroxide mist had something to do with it.

NHRA SANCTIONS ROCKET CARS

Ky Michaelson, owner of the *Pollution Packer*, gives the rocket car one more check before Dave Anderson blasts down the track.

The next NHRA National race was on March 18, 1973, at the Gatornationals down in Gainesville, Florida. Dave's first run at 305.80 miles per hour with an ET of 5.17 seconds set two new records—the fastest run and quickest ET ever on an NHRA track. While we were in Florida, we made a couple of exhibition runs at a track in Hollywood, setting two more records. After those two runs, I told Dave to vent the nitrogen out of the fuel tank so I could drain the fuel. I always did that as a precautionary measure so there was no chance of the hydrogen peroxide leaking onto the wooden floor of our semitruck. I knew if it did, there was a pretty good chance of it starting on fire. As it turned out, he inadvertently opened the wrong valve and, unbeknownst to me, as I got down on my knees to unscrew the drain plug, there was still 600 psi inside. As I began unthreading, the drain plug shot right out of my hand with such unbelievable force that it became buried in the ground. The hydrogen peroxide sprayed all over my face. Luckily, I acted on instinct and poured a 5-gallon bucket of water over my head, face, and hands before the burns got any worse. Thank God I was wearing polyester clothes, or I would have turned into a giant fireball. Cotton would have ignited because it's organic. I had severe burns on my hands, and my face looked like hamburger. I was so worried that if news of this accident got back to the NHRA officials, they'd likely ban rocket cars due to the danger of handling the fuel, so I immediately left the track with the team. That was a very painful experience but also a very valuable lesson, and I was lucky—my wounds eventually healed.

ROCKETMAN

The first 300-plus mile-per-hour run ever—at 305.80 miles per hour and an ET of 5.17 seconds—which again set the fastest and quickest run ever on an NHRA track.

Our next national event was on June 10, 1973, at the Springnationals in Columbus, Ohio, where we made quarter-mile history again. Dave ran the first four-second run ever, at 4.99 ET and 322.00 miles per hour. To put this into perspective, Top Fuel cars were running under 250 miles per hour within the mid 6-second bracket. The *Pollution Packer* looked more like a cruise missile than a dragster. After that run, a very famous drag racer came over to me and said, "Ky, thanks a lot for screwing up drag racing!" Thirty years later, the Top Fuel and Funny Car dragsters are achieving those speeds, which is absolutely phenomenal. In my mind, I knew we hadn't tapped into the full potential of that car. Next came the Nationals at Indianapolis, where we made history again with a 344.820-mile-per-hour, 4.62-ET run.

Dave Anderson drives the *Pollution Packer* at night.

NHRA SANCTIONS ROCKET CARS

As Dave continued to break records at tracks all across the country, things between Tony and me heated up. As good things often do, our rocket car collaboration eventually came to an end. In this case, it seemed Tony couldn't stand sharing the spotlight with me and somewhere along the line had decided his Pollution Packer company didn't just sponsor the car but actually owned it. In reality, I was still the owner and always had been. Tony tried pulling a major power play on me and actually had the semitruck driver hide the car after a race in Englishtown, New Jersey. That little stunt brought Tony a lawsuit, but rather than let the attorneys get all the money—and he had a lot more than I did—I decided to settle out of court. I entered into an agreement with Tony whereby he would pay me to build and install a new 3,500-pound-thrust engine that Dick Keller had designed for the *Pollution Packer*, and in return, he'd buy the car from me, but I would keep the original engine. The new rocket engine that Dick had designed was unique because it was cast out of stainless steel and then machined. That made it a lot easier to mass produce if we wanted to. Tony also paid me to go on tour with the car to train a new crew to run it properly.

With tension so thick you could cut the air, I managed to fulfill my obligations under the settlement agreement, but let me tell you, the second it was up, I moved out to California, where there was a lot of racing action, to build an all-new rocket car. I didn't have a lot of money, so I was fortunate enough to have a number of racing friends who allowed me to hang my hat at their places. I spent a lot of time with Jim Deist, owner of Deist Safety Company, manufacturers of parachutes, fire suits, and other safety equipment for racers. The one thing I liked about Jim was that he had the same two hobbies I did: racing and eating. We'd go to the local McDonald's about five times a week and have the ever-popular Egg McMuffins. Even after all these years, I *still* just love those things.

Chapter 6

Rocket Racers

Dave Anderson makes test runs in the *Pollution Packer Bonneville Dragster*.

In the spring of 1973, Ramón Alvarez and Russell Mendez had Glen Blakely build a rocket car for them. Arvil Porter built the propulsion system, and they named the car the *Free Spirit*. On July 6, 1973, Russell received NHRA license number 004 to become the fourth driver approved to drive a rocket car. The team made its first test runs at the Orlando Speed World track in Florida. During those test runs, the tires came off the front rims and, miraculously, at a cool 280 miles per hour, Russell somehow guided himself to a safe stop. He made a number of runs with the car on the East Coast, throughout the Midwest, and even once down in Mexico. I was fortunate enough to see Russell run at Bob Metzler's Jet Car Nationals one Memorial Day weekend in Union Grove, Wisconsin. I was amazed as I watched him lose control on his second run and go off the track and onto the grass, with absolutely no damage to the car or to himself. He later told me he wasn't sure what had happened but thought perhaps the crosswinds at the end of the bleachers may have been a factor in him losing control. He managed to make a 325-mile-per-hour run in 5.22 seconds ET. That was the last time I saw him run the car.

On one of my trips down to Florida, I stopped by their shop one day to say hello. They very eagerly explained to me some recent modifications they'd made in hopes of increasing the car's performance. One of the changes was to add wheel fairings to the front and rear wheels, which they thought would help streamline the car. The second I saw those fairings on the front wheels, I told Russell he had better take them off or they'd kill him.

He said something to the effect of, "Ky, you're just jealous because you don't have them on your car."

I replied, "No, Russell, there's a *reason* I don't have them on my car."

Russell Mendez drives his *Free Spirit* rocket car at the Great Lakes Dragaway in Union Grove, Wisconsin.

If you ever watch a rocket car make a run, you'll see that when the rocket motor shuts off, the car will always dart to the right or to the left. This is caused by a number of things, one of which is a change in center pressure; as soon as you remove the force from the rocket, the front end of the car suddenly becomes very light. The fairings on the *Free Spirit* weren't built right in the first place, as there was too much fairing in front of the kingpin, causing the car to oversteer.

Russell had one thing in mind and one thing only: to break our records. He showed up at the sixth annual NHRA Gatornationals on March 16, 1973, his birthday, with his girlfriend and family. He made an all-out run, and the wheel fairings did exactly what I knew they'd do. When he went through the quarter-mile, he lifted off the throttle, and the car appeared to veer slightly to the left as it crossed the finish line but then made an abrupt right turn into the guardrail. Russell was thrown out of the car and killed instantly. That was a very sad and tragic day as the rocket racing world lost a true pioneer.

Back when I worked at Tony Fox's factory, I worked along with Tim Kollach on a monocoque car called the *Bonneville Dragster*, which had been designed by Dick Keller. Tony's team transformed it into a second *Pollution Packer* and dubbed it the *Pollution Packer Bonneville Dragster*. I had seen Dave Anderson make a few runs with the car, but it seemed to have handling problems on its initial test runs. The team continued making exhibition runs around the country with it and set a number of track records.

Even though I was no longer a part of the team, I remained very interested in what the original *Pollution Packer* was doing, and I distinctly recall being at George Weplow's race car shop one day when I asked him if he knew where Dave was running the following weekend. George told me he was running the old car at Charlotte, and I actually said, "If he runs on that track, he'll be killed." I made the statement because I knew the track was an oval race course with a 1/8th-mile drag strip that went through the middle down pit row. Dave had never run on a 1/8th-mile track before, and I knew how fast the *Pollution Packer* was. He made his exhibition run at Charlotte Motor Speedway, and he went all out; sadly, it turned to tragedy, and my prediction came true.

I watched the video of the run some time later, and it made me sick. I could plainly see that the *Pollution Packer* was traveling 248 miles per hour, which was much too fast to stop on such a short course. That accident really took its toll on me personally, as I watched my lifelong friend lose his life, along with two other innocent people who were working on their car at the end of the track. Dave was a first-class gentleman who died doing what he loved to do and was respected by everyone who knew him—another huge loss to the world of rocket car racing.

A few months after Dave passed away, Vern Anderson, no relation to Dave, was given NHRA rocket license number 007 to drive the *Pollution Packer Bonneville Boss* car. His best time on a drag strip was accomplished on August 11, 1974, at Albuquerque, New Mexico, where he ran 344.820 miles per hour in 4.62 seconds.

Chapter 7

John Paxson

John Paxson on the return road after making a run in his new *Armor All Rocket* dragster at the Great Lakes Dragaway in Union Grove, Wisconsin.

News that Bill Fredericks had built an all-new rocket motor for the *Courage of Australia* with a reported 8,500 pounds of thrust, replacing the original 3,000-pound motor, soon traveled throughout the rocket-racing circuit. NHRA official Bernie Partridge became extremely interested as well. We all knew that if this were in fact true, it would be the most powerful rocket motor ever used in a quarter-mile car.

The NHRA immediately ordered a recertification run for the car, so Bill and driver, John Paxson, brought it out to Irwindale Raceway in Southern California. John made an easy first run and shut the motor off at the 950-foot mark, coasting through the lights at 250 miles per hour in 5.52 seconds. Unfortunately, disaster struck as he hit the main chute: it didn't deploy. He immediately hit the backup chute, but it didn't deploy either. He hit the brakes so hard the wheels locked up, causing the narrow Goodyear land speed tires to burn right through. With no chutes and no braking power, the car entered the sand traps at a high rate of speed and hit an embankment, causing it to catapult through the air. It finally landed upside down in a dry riverbed, and the emergency system kicked into action, venting the hydrogen peroxide fuel, which sprayed up in the air. The chutes hadn't deployed because the new rocket engine was much longer and actually stuck out of the back of the car. When John attempted to deploy the main chute, the pop chute stuck to the hot 1,300-degree rocket motor chamber, and as he hit the reserve parachute, its pop chute also stuck. Even though the fuel vented, the whole scene could have easily been a disaster. Luckily, there was no major fire, and John wasn't hurt. The car, on the other hand, was completely destroyed.

Not long after that crash, John teamed up with Steve Evans, Bill Doner, and Grant McNiff to build another car. M and S Racecars in Azusa, California, built the 21-foot chassis, and Bill Fredericks installed the rocket engine from the *Courage of Australia*. The car cost around $30,000 to build and featured Cragar wheels, Airheart brakes, and Deist chutes. The team picked up a sponsorship from Armor All and aptly named this new creation the *Armor All Rocket*, which debuted in June 1973 at Sacramento Raceway.

John received NHRA license number 005 for the car and, two weeks after its debut, made a 300-mile-per-hour pass at Irwindale Raceway, followed by a record-breaking 326.70 miles per hour in 4.82 seconds run at Albuquerque, New Mexico, a week later. The team was on fire and actually picked up a second major sponsor, the Revell Toy Company. They had over 30 bookings in six months before John raced up in Canada, where a slight design change was made. He decided to have Bill Fredericks install a nitrogen gauge in the driver's compartment as an added safety feature. His former crew member had always set the pressure through the dome loader mounted behind the roll cage, and he wanted the new team to have this same access to avoid over-pressurizing the tank. Unbeknownst to anyone, when Bill drilled the hole into the stainless-steel fitting to mount the gauge, a small piece of it got into the nitrogen pressurization system and ended up getting under the seat of the regulator. No one was aware of this at the time and had no idea it would prevent the regulator from controlling the nitrogen pressure. It wasn't until John was on the starting line, getting ready to make a run, that the fuel tank suddenly overpressurized to 1,200 psi, causing the safety burst disc to rupture. The hydrogen peroxide shot out of the tank, hitting two of his crew members, who were both standing next to the car and were unfortunately wearing cotton Levi's instead of polyester, and their pants immediately

JOHN PAXSON

Brad Proffitt in his *USA-1* rocket dragster, getting ready to make another high-speed run.

caught on fire. Luckily, they had water nearby, or things could have been a lot worse. One of the most horrifying things about the entire experience was that no one knew exactly how many people had been sprayed, and there were at least 10 whose shoes caught on fire from the incident. John relayed to me how fortunate they all were that things hadn't been a whole lot worse.

Danger and mystique for this team seemed to lurk around every corner, including the time John was at a track in Ohio with photographer friend and crewman, George Callaway. John wanted George to get a picture of the car with a lot of smoke coming out the back, so he purposely didn't warm up the catalyst pack before the run. When he hit the throttle, the majority of the peroxide didn't decompose, leaving a *huge* white cloud of vapor. When he got out of the car at the end of the track, it was so thick and heavy in the air that he couldn't even see the tower, and his tow car driver had to wait for it to disappear before he could drive down to pick him up. When they got back to the pits, they found out they had set several small grass fires and that a number of spectators had suffered bleached skin from the peroxide. In addition, the shoes of some of the people working on the starting line caught fire when they stepped into the raw material. Even though things like this happened periodically, John told me he made close to 300 runs in the car with very few incidents. In my mind, he was definitely a first-class guy who was always safety minded, but as the old saying goes, "Sometimes you eat the bear, and sometimes the bear eats you."

The car was later sold to Ray Alley, who changed the name to *Age of Aquarius*. Ray campaigned the car for a short period of time, and his best run was 325 miles per hour with an ET of 4.88 seconds. In 1978, the car changed hands again when it was sold to Hayden and Brad Proffitt, NHRA license number 032, and renamed *USA-1*. Brad made over 200 runs until he retired it in the early 1980s. John actually bought back the car, had it restored, and donated it to the Wally Parks NHRA Motorsports Museum in Pomona, California.

The legendary Budweiser land speed car on display.

BILL FREDERICKS

Bill Fredericks designed and built the magnificent ***Budweiser Rocket Car***, which was truly a masterpiece. He built a hydrogen peroxide hybrid rocket motor for it, using a number of solid rings of polybutadiene plastic inside. In simple terms, the hydrogen peroxide goes through the catalyst pack and decomposes, generating superheated steam and oxygen at temperatures of 1,370 degrees. The high temperatures actually erode the solid fuel (polybutadiene), which then automatically ignites as soon as it's gaseous. Fredericks also used a sidewinder missile motor for an additional 6,000 pounds of thrust. Stuntman Stan Barrett drove the car and, on December 17, 1979, set an unofficial one-way speed record of 739.67 miles per hour. Unfortunately, for some reason, Bill had no desire to go after the official two-way land speed record, which, in my opinion, the car could have easily broken. Even today, a cloud of controversy surrounds their effort.

Chapter 8

Paula Murphy

Paula Murphy stands by Ky Michaelson's STP rocket car.

In 1973, I built a new rocket car. It was much improved over the *Pollution Packer* that Paula Murphy had driven for me in the past, because I laid the fuel tank down horizontally and placed the nitrogen bottles further forward, behind the cockpit. It seemed like I no sooner got the car built than I received a sponsorship from the STP Corporation. I was fortunate enough to commission Paula to drive for me, so we named it the *Miss STP Rocket Dragster*.

Paula, who at that point in her career was known as "Miss STP," was a true pioneer in the world of auto racing. She had her first taste of the sport while attending Bowling Green University in Ohio, where she won the state K-class catboat racing title. Racing was the furthest thing from her mind when she moved to California and took on a secretarial job, but that changed the day she bought an MG sports car as her means of transportation. Driving the MG prompted her to join a local sports car club, which exposed her to Southern California road racing.

It wasn't long after attending a number of races that Paula was offered the chance to drive an Alfa Romeo race car in a women's race at Pomona. Much to her surprise, she won, and that event led to a seven-year career in sports car road racing. She wound up a professional driver, racing Ferraris, Aston Martins, Maseratis, and Porsche Spyders.

In 1961, she drove the Mobil Economy Run, a performance that led to a job as a test driver for Andy Granatelli, who was then performance director for the Studebaker Corporation. While driving at Bonneville for Granatelli, Paula set a new world flying-mile speed record for women of 161.29 miles per hour in a Studebaker Avanti, plus a series of speed and endurance records.

In 1963, Paula Murphy became the first woman ever allowed to drive a race car around the 2.5-mile oval at Indianapolis Motor Speedway, and she toured several laps in the STP-Novi at better than 100 miles per hour. By 1964, she was driving Walt Arfons' *Avenger* jet car, which led her to a two-way world record for women of 226.37 miles per hour, including one run of 243.44 miles per hour. She went down in automotive history for being the first woman to top 200 miles per hour in a race car.

The author works on the STP rocket car.

Paula decided to switch from sports car racing to drag racing because of the former's lack of prize money available to professional women drivers. Her first drag race was in a Super Stock Oldsmobile, in which she set a national record. She literally blazed the trail for the women who followed by fighting all the major political battles of motor racing when there was no women's movement. Paul demanded and got the very first nitro-burning race car license ever to be issued to a woman by both the NHRA and the American Hot Rod Association (AHRA) and, on August 24, 1969, became the first woman ever to hit 200 miles per hour in the quarter-mile in her STP-sponsored Funny Car while qualifying for the AHRA Nationals. As I mentioned earlier, Miss Murphy was the third driver and the first woman to receive a competition NHRA rocket license (number 003). She went on to break new ground for women in racing again when she set a standing-start quarter-mile two-way record of 6.708 seconds in October 1972 at the Bonneville Salt Flats in my *Pollution Packer* rocket car.

PAULA MURPHY

Paula Murphy in the STP rocket car, ready to make an exhibition run at Sears Point Raceway.

Needless to say, I was ecstatic to be working with her again, but as we headed to northern California with the *Miss STP Rocket Car*, where Paula was scheduled to make two exhibition runs at Sears Point Raceway, we did not know things were about to go horribly wrong. For some reason, at the end of the first run, she hit the main chute and the reserve chute simultaneously, leaving her virtually brakeless as the force of the chutes' drag literally ripped away the portion of the frame securing the chute tow lines. The g-forces knocked her out cold, and she crashed hard. Paula wound up cracking both her third and fifth cervical vertebrae in the collision, and the car was completely destroyed. My personal thoughts on how the crash occurred are based on Paula's previous driving habits. She was a very experienced Funny Car driver, and at the end of her quarter-mile runs in her Funny Car, she would always push the clutch in with her left foot and deploy her parachute with her right hand. The *Miss STP Rocket Car* had a parachute lever by her right hand and another by her left foot, so I think she just instinctively pushed the lever, as she had done hundreds of times before in her Funny Car. We were truly blessed that she came away from that accident with injuries as minor as they were, because it could have been a whole lot worse. The car I could replace, but Paula? No way.

Shortly after that crash, I went back to Tarzana, California, where I spent a lot of my time in Frank Huzar's shop building another new car. Paula and I had become very good friends, so she put me up at her home, where she lived with her son and her 90-year-old father, who I thought was one of the coolest guys I'd ever met. Due to her injuries, Paula had to wear a full upper-body cast to support her neck. Even though she definitely looked a mess, it didn't stop us from having a great time running around together and going to the drag races at Orange County Raceway. She was a real go-getter and, as soon as the cast was removed, she began touring the show circuit during the winter months. It was while doing so that she made a very big decision. She called me from a World of Wheels show in Kansas to let me know she'd decided to retire from car racing altogether so she could spend more time with her son and her father. I understood her reasoning completely, and her decision prompted me to head back to my roots in Minnesota, where I hadn't been for quite some time. I packed up my things and brought the car back with me.

Jerry Hehn asked me to build him a new 3,500-pound rocket motor and propulsion system for the car he had just built called *The American Dream*. I was thrilled at the opportunity and was fortunate enough to work with Jerry for as long as it took him to become comfortable in the car and with the whole rocket-racing scene—which wasn't really very long but was definitely a great experience for me nonetheless. Jerry made his qualifying runs at my hometown track, Minnesota Dragways, and in no time at all held NHRA rocket license number 008.

The track rescue crew looks over the wreckage of the STP rocket dragster.

Chapter 9

Ed Ballinger and the *Conklin Comet*

I received a call from Ed Ballinger, who I had met at an aerodynamics seminar in California, and he told me he'd been working with Craig Breedlove throughout the years. The first time I met Ed, I knew he not only had a great personality but also a lot of mechanical ability. All in all, he was a very likeable guy. Well, he'd heard about the car I'd built while in Tarzana and wanted to know if I happened to be looking for a driver. I told him his timing was impeccable, because not only was I looking for a driver, but I was also looking for a sponsor. I told Ed the only name on the car was mine and we would need money for hydrogen peroxide and a nice tow rig so we could travel in style. Our conversation ended in an agreement that he'd partner as my driver and we'd work on finding a sponsorship together. The only downfall for poor Ed was that I somehow convinced him he'd have to move to Minnesota despite the cold weather. I knew if he'd truly had his way, we'd have moved my entire racing operation back out to California, but I was resettled and wasn't up for another hiatus.

Ed arrived and got all squared away. The first thing we needed to do was to get Ed in the car and get a few runs under his belt, so we went out to Minnesota Dragways on a weekday to get some practice runs in. I found out that Ed was a very cautious driver when, after his first run, he said the famous phrase, "There are old pilots and bold pilots, but there are no old, bold pilots." He was in no hurry to break any speed records and, in my mind, that was good. Ed continued to make a number of test runs, and I became convinced he was a capable driver and was ready to get his rocket driver's license. I got on the horn to Bernie Partridge at NHRA headquarters and asked if he'd come up to Minnesota and license Ed. He agreed to make the trip and within a few days was out at the Dragways to watch. It only took

Ky straps Ed into the car before he makes his license test runs.

a couple of runs, and Ed was issued NHRA rocket license number 010.

With that out of the way, our next challenge was to find a sponsor. Ed was thumbing through the phone book one day and saw an ad for an advertising agency called Adams and Others. The name intrigued us, so Ed called and arranged a meeting with a Mr. Milt Adams. We told him we had a rocket car that could get one of his clients a lot of publicity. He seemed very certain he could find someone to back us and, sure enough, a few short days later, we got the call we'd dreamed of. He signed on the Conklin Company as our sponsor and they agreed to back us to the tune of $50,000, which was a *ton* of money back in those days.

Let me tell ya', we were like two kids in a candy store. The first thing we did was lease ourselves a motor home in order to haul our race car trailer. Because we thought it would be quite a challenge to try to set a land speed record in every state in the country, we knew we'd be traveling extensively. Next up was the car. Now that we had a sponsor, we could give it some character, so we ended up building an all-new streamlined body that we painted in multiple candy-apple colors, but that's not all: we even went so far as to have the roll cage gold-plated.

Our new car, which we proudly dubbed the *Conklin Comet*, was absolutely drop-dead gorgeous, to say the least.

At just about every track we ran on, we set a new elapsed time and state speed record. We raced from one end of the country to the other, starting out in March and going all the way through October. During the winter months, Ed chose to stay down in the southern states and pick up a few racing dates while there.

It wasn't enough that he got the races all lined up; he also felt the need to promote himself and the car as well. On just such an occasion, the local track he was going to run on asked him to do a promotional spot, so he did a radio show. No big deal, just a normal part of being on the circuit. Well, he had no idea that at the time he did his piece, the drag strip was having problems with homeowners who had chosen to build right next to the strip. They claimed that the noise was harming the environment and the wildlife; needless to say, when Ed announced he was going to run a 300-mile-per-hour rocket car at the track, all hell broke loose. They even went so far as to try getting an injunction against the track to prevent him from running. While they failed at their attempt, a number of people from the Environmental Protection Agency showed up with decibel meters in hand to measure the sound of the rocket. While Ed warmed up the engine, the *Conklin Comet* nearly blew down the fence directly behind him—that was warm-up. He succeeded shortly thereafter as he made his run down the track: the rocket completely destroyed the fence, as pieces of wood flew everywhere. As Ed exited the car in the pits afterward, two of the EPA people came over and were actually laughing hysterically.

One of them had gone so far as to put a piece of wood from the fence in his mouth and then jokingly said, "This is ridiculous. Look what happened to me from just standing there."

It seemed those guys were so impressed with the performance of the car that they forgot all about the noise factor. They shook Ed's hand and left the track together, still laughing.

Ed and I weren't alone in our traveling adventures. Every so often, while they were on school vacation, my sons, Curt and Mike, used to hit the road with us and help us out. Curt was 14, and Mike was 16. Curt always expressed the desire to drive the rocket car when he got older, but I had different ideas because of the dangers of high-speed racing.

The *Conklin Comet* on the starting line versus the *Pollution Packer*.

Chapter 10

Captain Roller Ball

Speaking of other ideas, I was at a scuba shop one day when an idea hit me like a lightning bolt to the head. I noticed this aluminum backpack in the corner, being used to hold scuba tanks. I immediately thought to myself, "Hey, what if a guy was to build a hydrogen peroxide rocket backpack?"

Don't ask me why that question popped into my head, because what I'd ever do with it, I really had no idea. I decided to build it anyway, before somebody else did. Besides, it would be a nice break since it was the end of

CAPTAIN ROLLER BALL

Ky tests his rocket backpack in front of his house in February 1973.

the racing season. I didn't let on to anyone what my latest little project out in the shop was, and I wasted no time in starting on it. Because of the high heat of decomposing hydrogen peroxide and the weight of steel, it was necessary for me to use stainless steel, so I machined the small 250-pound-thrust rocket from that and then used an aerospace high-pressure vessel for the nitrogen tank. That was a good beginning, but I needed aerospace hardware that wasn't readily available in the Twin Cities area, so I flew out to California to several surplus shops I knew. Incredibly, it was actually cheaper for me to make the trip than it would have been for me to buy the items I needed brand new. After rummaging through several places, I found all the components I had been looking for and headed back home to complete my latest invention. It ended up taking me about three weeks to get the backpack built and, by the time I finished, we were in the middle of a very long, cold Minnesota winter. There I was: all excited to test this thing out, and it had to be cold and icy.

    I decided I wasn't gonna let that stop me, though, so I got on the horn to Ed and another friend, T. J. Snow, and asked them both to come over and give me a hand firing the rocket. Let me tell you that even a true Minnesotan doesn't come out of hibernation without good reason, so I gave them both just enough of a description to pique their interests, and it worked. Ed happened to be in Minnesota, so both men arrived in a very short time, and we headed out to my shop for the big unveiling. Of course, as usual, I had a plan: I'd strap the backpack to the telephone pole out in front of my house for its first test. Just as we were doing so, Ed says to me, "What's the matter, Ky? Don't you trust your own workmanship? Are you scared of it or what?"

That little challenge caused me to change my plan, and without further hesitation, I told him to just strap it on me directly. He, of course, thought I was crazy, but he did what I asked. Strutting like a peacock, I walked from the telephone pole over to the long stretch of sidewalk out front that just so happened to have a real thin coat of ice on it. Traffic was cruising right along and there I was, a grown man standing on an icy sidewalk in the dead of winter with this never-before-seen rocket-powered backpack strapped onto my back.

Curt Michaelson on Penn Lake, getting ready to blast off on ice skates with the rocket backpack.

I said, "Okay, here goes nothing, guys," and I slowly opened the hand-held throttle valve. The hydrogen peroxide flowed into the rocket chamber, and all of a sudden this ear-piercing sound came out of the nozzle, and the rocket came to life and started pushing me down the sidewalk. It felt like a giant man pushing on my back. I had to crouch down so it wouldn't push me over. Those guys started screaming with laughter. I couldn't *believe* how much power this thing had and how well it worked! I was gone. That was it. One of the most awesome little rockets ever made, and I had invented it, and it was *fun*. I knew if I thought this was fun, my son, Curt, a natural hockey player and darn good skater, was going to go absolutely nuts when he got a load of it, so I literally ran into the house to get him to come out and check it out. He was game, no problem. Instead of using the sidewalk again, we decided to give this thing a try on something a lot more fun—the small, frozen lake out back. Curt tied on his hockey skates, and we strapped the backpack snugly onto his back. I showed him how to work the throttle and gave him some last-minute instructions, and he was ready. He actually added his own twist to the performance by first skating as fast as he could, and then he crouched down and slowly opened up the throttle.

Curt Michaelson and the author check out the rocket backpack before Curt's test runs in his school's parking lot.

I'm telling you, he took off like a human rocket, and he actually made it all the way across the lake before he ran out of rocket fuel. I had never seen anyone move that fast on a pair of skates, and I thought to myself how the neighbors on the other side must have been wondering what kind of an unknown speed skater lived in the house across the way and how the heck he had just done what he did. He looked like a cannon ball in flight. It was totally cool—really, really loud, and I was in awe standing there watching my son having the time of his life. He skated back with the biggest smile on his face and excitedly announced to me, "That was *awesome*. Hey, Dad, I want to do this on roller skates at drag strips!"

Aha! What a terrific idea. Curt really had something there, and he looked really good while doing it. Ed and T. J. agreed, so I told Curt I'd definitely check into it. We got everything packed away in the shop, and I thanked those guys up and down for all their help. That was a very memorable, cold winter's day, to say the least.

This whole new entertainment idea came about prior to expanding my Space Age Racing Inc. business, so I still had a booking agent whom I was paying. He was a good guy by the name of Duane Nichols, so as soon as I got back inside, I headed right into my office and called him. I told him what we had and asked whether or not he thought it would be possible to add Curt to our bookings at the same drag strips where we ran our rocket car. He said he'd see what he could do and he'd let me know. Curt was as anxious as I was to hear back from Duane, and it didn't take long. He called me back the next day with great news: he had a number of track owners who wanted us right away, sight unseen.

I'll never forget the look on Curt's face when I told him he was now going to become the famous "Captain Roller Ball." He was in heaven, and neither of us could wait for winter to end. It was one of the longest winters I can remember, but we stayed busy preparing. Curt took the opportunity very seriously, and he spent many hours practicing skating with the 40-pound backpack on his back until he learned how to balance completely with all that extra weight.

As soon as the snow melted, he practiced in the school parking lot, making slow-speed runs with his school buddies looking on. It wasn't long until he made a high-speed run for his teacher and all of his classmates, and they all thought he was really something. I couldn't have been happier for him, because I could clearly see how much enjoyment he took in this very unique sport of his, and I knew he'd become somewhat of a celebrity because it was just something so different. As Curt practiced, I had my friend, Jim Deist, sew up a yellow-and-blue cape and a flashy silver fire suit, just like the Top Fuel racers use. This was back in the early 1970s, so there was no such thing as in-line skates back then. He had to use the old-fashioned, four-wheeled roller skates and, for night appearances, I fastened flashlights underneath them so he could see any debris that might be in his path.

His first professional appearance was at the annual Oktoberfest celebration over in La Crosse, Wisconsin. The racetrack is fairly large, and the event is well known and attended by thousands of people. When we arrived, there was a grandstand *full* of 'festers, and he was fighting chilly nerves on a very chilly October day. I gave him a pep talk and got him as fired up as possible, and he hit the track, uncertain of what lay ahead. He had his fire suit, cape, and helmet on, and the crowd was totally into it. They weren't sure what exactly they were looking at, but it definitely looked interesting, and they were behind him all the way. He took his position, started to skate, and when he hit the throttle, he was clocked in at 30 miles per hour on his first run. He received his very first standing ovation from the crowd that crisp, cold

> "There's not a man in the world I wouldn't fight, but there's no way I'd ever go on rocket roller skates."
> —Jerry Quarry

CAPTAIN ROLLER BALL

October day and from then on, he *was* Captain Roller Ball, the showman. After that, no matter where we traveled, the rocket backpack and my son were the talk of the town. People found Curt's stunt to be more than a little odd, but very, very fascinating. It wasn't long before he earned his place in the record books with speeds in excess of 50 miles per hour. The best part of the whole deal for him was that he started earning $350 per appearance, which was a heck of a lot of money for a 14-year-old kid. He also started receiving a lot of national attention through magazines, newspapers, and television—news of Captain Roller Ball traveled fast.

It just so happened we were racing down in the New York area when we received an invitation to appear on *The Mike Douglas Show*, which back then was the equivalent of *The Tonight Show with Johnny Carson*. Since we were in the area anyway, and this was a wonderful opportunity, we accepted the offer and headed to Philadelphia for the taping.

We were thrilled at meeting celebrities and were honored to learn we'd be appearing with Mike Connors and comedian David Brenner, who had also volunteered to try out the skates. Curt made his run down the alley in back of the studio, where a couple of guys were holding a mattress at the end of the block to stop him from flying through the intersection. As always, he made it look easy.

Next came David, who became a comedian the second we strapped on the backpack, asking how far it was to the Manhattan toll bridge, to which Mike replied by handing him 75 cents. He started jumping up and down on his skates and then opened the hand throttle just enough to let a little peroxide flow into the chamber. The rocket just went "pop, pop, pop," so he started making fun of the lack of power. Everyone was laughing, and that's when he made the mistake of really showing off and opened the rocket throttle wide open. It picked him up and threw him right on his face. He hit the ground pretty hard, and at first I thought he was hurt. We all ran to him, and he looked up and asked if we were there yet. Everyone broke out laughing, including David. The show turned out to be a huge success and an experience that none of us will ever forget.

David Brenner wasn't the only one who tried out the backpack. There were a few occasions when I let stuntman Super Joe Einhorn give it a try. In fact, we wound up working together on a live ABC show called "Superstars," where he was scheduled to try to break Curt's record on the show. Unbelievably, the promoters got worried that the stunt was too dangerous and if something bad happened, they'd be held responsible. Super Joe told them and the crowd that had gathered not to worry; that he'd accept full responsibility if anything happened. As he strapped into the backpack, a member of the production company, Barry Frank, blurted out, "This guy doesn't have any insurance or any assets." Then Jerry Quarry, who was maybe the most fearless boxer alive at the time, chimed in with, "There's not a man in the world I wouldn't fight, but there's no way I'd ever go on rocket roller skates." Needless to say, Super Joe was not allowed to rocket into the record books on the show that day, and Curt went back to being the only one to make appearances. He continued to appear at racetracks for the next couple of years, thrilling the crowds with his high-speed antics. He was never injured, never fell, and will go down in history as the first rocket-powered roller skater, setting the speed record at 52 miles per hour in 1976. I'm proud to say that Captain Roller Ball lives on.

Chapter 11

Space Age Racing Inc.

Ky proudly displays a few of the powerful hydrogen peroxide–powered motors he built in his Bloomington, Minnesota, shop.

Somewhere between knowing and not knowing what I truly wanted to do with my life, I signed up for a machine-shop night course at St. Paul Vo-Tech. During my very first class, the shop teacher said something to me that would truly change the course of my life.

He said, "Ky, I don't care what you make here as long as you learn."

I was ecstatic. The course only cost me $38, and I had access to one million dollars' worth of machinery. Since he had told me I could build anything I wanted, I chose rocket motors. That's when it hit me like a rock: I had free-and-clear access to all this equipment, and I could whip together a motor in no time, so I took out an ad in *National Dragster* magazine that read, "Save Nitro—Burn Hydrogen Peroxide." Much to my surprise, piles of orders came in. So many, in fact, that I started a company called Space Age Racing Inc. and actually made over $50,000 just in the time I was in school.

Now *that's* what I call a good educational program. In no time, I was able to expand the business. In addition to the motors, I began building rocket-powered vehicles for other people nearly every winter, not only building cars, but I got the bookings for them as well.

In return for my services, the vehicles' owners would pay me 15 percent of their gross profits. The one thing I failed to realize back then, though, was that every time I built a rocket motor for someone else, I was only creating a new competitor who could potentially take away my bookings.

I built the first rocket-powered Funny Car in my shop in Minneapolis for Jim Hodges in 1973, and he called it the *Alabama Express*. Jim raced Funny Cars for many years in the southern United States and was a very nice guy with a thick, southern drawl.

I used to call him the Gator Man. When he wasn't racing cars, he'd take his race car trailer down to New Orleans, fill it up with shrimp and ice, and bring it up to Minneapolis to sell to the local seafood stores. I remember one time he left his trailer parked across the street from my house a lot longer than he had planned to. After about a week, every cat in town was trying to figure out how to get inside, while the rest of us had to put up with the overwhelming odor.

A Ky Michaelson–built hydrogen peroxide–powered motor bolted to the chassis of Jim Hodges' rocket Funny Car.

Chapter 12

Side-by-Side Go Karts

Shortly after a series of appearances, I decided I was tired of paying someone else to line up shows for us, so I expanded my business to become a booking agent for others. The timing couldn't have been better because it was also when, thanks to a visit from my friend Pat Best and his friend George Lavigne, we decided to step things up a notch. They came over to my house to talk to me about building a couple of rocket-powered go karts because they had seen Captain Jack McClure's rocket-powered go kart in a magazine. Jack had driven to an unbelievable speed of 200 miles per hour in 6 seconds. I had actually seen him run a number of times, and he no doubt put on one heck of a good show.

Well, these two thought they could do one better and told me they wanted two karts so they could race side by side. My first thought was that they were crazy, but they went on to tell me they were dead serious and that while their show would appear to truly be an all-out race, their plan was to flip a coin beforehand to choose the winner. That way, it wouldn't be so dangerous. They were really only interested in putting on an entertaining show and getting paid for it. I thought it was a great and exciting idea, so I agreed to build the karts under one condition: I would do so with a fuel tank small enough to only allow them to go 150 miles per hour in the quarter-mile. In my mind, going side by side, that was definitely fast enough. They agreed, so I set to work and got Pat's done, but because we had to get ready for the next racing season, I was unable to finish George's. Since they had just bought a new Ford cube van to haul their racing go karts together, George decided to come to California with me so I could finish his from there.

We'd made the trip from Minnesota to California many, many times, so once again, Ed, T. J., and I packed up, and George gave me the first shift at

George Lavigne's *Thunderbolt* go kart with small fuel tank.

driving the new Ford van. We hooked up my big enclosed trailer, with my rocket car and the go kart shell inside, and headed on down the road. Ed and T. J. drove the Jeep Wagoneer for the first couple hundred miles, and when we were about halfway through the state of Iowa, we stopped to get a cup of coffee and something to eat. As we sat in the restaurant, I looked out the window and noticed it was starting to snow pretty heavily. I mentioned that maybe we should get a motel and wait for the snow to stop and go on in the morning, but everyone else wanted to get to California to soak up some of that sunshine.

As we left, Ed volunteered to drive the van so I could drive the Jeep, and we headed off once again. We'd only driven for about an hour when suddenly the snow had gone from just a storm to a blinding, whiteout blizzard. It was so bad that I could barely see the taillights of my trailer in front of me, so I called Ed on the CB and told him we had better exit off the freeway as soon as possible. He agreed, but when we attempted to get off at the next exit, it was completely blocked by a huge snowdrift. There was no way to get through it, so we had no choice but to forge ahead. By that time, the freeway was down to one lane, and even the snowplows had abandoned the roadway and stopped plowing. Things were definitely not looking good for us, but we had no choice but to continue on. I looked in my rearview mirror and caught a glimpse of fast-approaching headlights that appeared to be a semitruck barreling down on me, and before I could do anything at all, the unthinkable

Captain Jack McClure blasts down the track on his rocket-powered go kart.

happened. It was a semi all right, and he had evidently not seen me, because he slammed his rig right into the back end of the Jeep and spun me sideways. I started screaming as he pushed me sideways down the freeway, with no way to break free.

I looked up in horror and could see the driver's face. He was driving with one hand and pulling on his air horn with the other, and then everything went blank. It was like a slow-motion nightmare. The Jeep had rolled over, and I was thrown out. How, I don't remember, but what I *do* remember is lying in a cold snowbank with major chest injuries and a hole punched in my head, suffering a severe brain concussion. There was blood coming out of my mouth, nose, and ears, and I was drifting in and out of consciousness.

I didn't know what happened to the semi, either, and recall a short time later someone coming and putting a blanket over me. From then on, things started to get really weird. It was like I was floating in the air looking down on myself and wondering why I was lying in the cold snowdrift with people standing all around me. Ed told me later they wondered what had happened to me and had to drive for over an hour before they could find an exit ramp that wasn't snowed in. When they came upon the scene of the accident, they thought I was dead, but the ambulance arrived and rushed me to the hospital. I guess it was actually a good thing that it was so cold, because it had helped to slow the bleeding. They stitched me up and took care of my chest injuries, and I ended up staying there for about a week. I was certain there would be no racing season for me after that, so even though I was in total misery, I told everyone I just wanted to go home and to get me out of that hospital.

They released me a short time later, and I flew back to Minnesota, where I received additional treatment by local doctors and got in some much-needed rest. I laid in bed for a few days, thinking of the guys out on the road, soaking up the California sunshine, and I became bored to tears. That's when I knew I had to get back to work, so I booked the earliest flight I could and headed back out.

I boarded the plane and slept for most of the three-and-one-half-hour flight. I awoke to a severe headache, and my ribs were absolutely killing me. All the pain pills the doctors had given me had worn off, and I had no more. When I stood up to exit the plane, I couldn't even walk straight. I'm sure all my fellow passengers assumed I was drunk. I wasn't sure what was happening, but I felt I had lost my equilibrium entirely. I managed to get out into the airport and was supposed to meet Ed in the baggage claim area. As I headed in that direction, I couldn't keep my balance, and it was a really long walk down the corridor. I had to keep putting my hand on the wall to continue going straight. I barely made it, and as soon as I spotted Ed, he rushed over, and I told him I was really messed up and needed a place to stay. Ed rented an apartment, and it took me a good week before I finally mustered the strength to work on my car again.

Once my strength returned, I went full barrel and somehow managed to get George's go kart done for him just a few short weeks later. He decided to call it *Thunderbolt*, which I thought was a great name. He had the kart painted metallic blue with white lettering and had Jim Deist make him a silver driving suit to protect him from fire and hydrogen peroxide. For added protection, Jim built a parachute for the kart itself, but a second one was also strapped onto George's back. That way, the kart would stop, and the secondary chute would also deploy in case he flew out, helping to avoid possible injury. He was as ready to go as he was gonna be.

Meanwhile, Ed and T. J. went on tour with the *Conklin Comet*, and I'd fly out on the weekends to run it. I no longer wanted to be on the road for six months out of the year, and they both understood completely. One weekend, though, because of two bookings we had in California, I knew I had no choice but to do a road trip. We had one in Orange County on Saturday night and one in Fresno, about 400 miles away, on Sunday night, so I knew it'd be an all-nighter to get to Fresno on time.

SIDE-BY-SIDE GO KARTS

We arrived in Orange County and were stunned by the huge number of rocket and jet cars they had scheduled to appear. We made our initial run but realized it was gonna be a long while before we'd make our second run, and it was. As the night went on, fog from the ocean began rolling in and, as luck would have it, we were the last ones to make our run before they actually closed down the track. At the time of Ed's run, the fog was so dense that as the car sped off, it appeared to follow him, as if pushing him down the track. I was in the tow vehicle and had to wait for the smoke to clear before I could head down after him. I got to what I figured was about halfway down the track and had to slam on the brakes. The fog was so thick at that point that I could no longer see past the hood of the truck and was afraid I would run right over the rocket car. I knew Ed was still somewhere on the track, but I had no idea where. By that time, I was going at a snail's pace, and as I crawled along in the blinding fog, I shouted out my open window, "Ed, where are you?" As I frantically searched, the fact that we had to be in Fresno in the morning for the next race was also in the back of my mind, and I was not a happy camper at the thought of another horrifying drive.

As bad as things were for us right then, we were not alone. Because the promoters had waited so long to close down the event, there were literally thousands of spectators now walking in the fog, too, searching for their cars throughout the parking lot. I finally found Ed, and it took us at least 20 minutes to hook up and tow our car back to our truck and trailer so we

George Lavigne with his rocket-powered go kart.

Ky, Ed, and T. J. Snow stand next to the *Conklin Comet*.

could load up and leave. We fumbled around, rushing to get on the road, but to no avail. There were so many spectators lost that nobody could get out of the parking lot. It was like feeling your way around in the dark. It was definitely the biggest mess I ever saw, or rather *felt*, my way through. We finally got out of there at about 2:30 in the morning, had no sleep, and started heading north to Fresno. The good thing about it was all the other cars that were booked in were coming from the same place in Orange County, so none of us arrived in Fresno until at least noon. Once there, our car made two runs, both over 300 miles per hour. The ride home after those appearances was one of the longest ever; everybody, and at times maybe even the driver, was sleeping.

By that time, my booking agency had really taken off, and I was setting up race dates for lots of the guys; they were happy, and I was happy. I was keeping everyone pretty busy, and the side-by-side go kart show was going really well. That is, until there was a personal conflict between one of the guys' girlfriends and his partner. Things got so out of hand that at the end of the season, the team split up. They decided to get their own trucks and travel separately but still race against each other. I wasn't sure how this would turn out, but I agreed to continue to book their events for them.

SIDE-BY-SIDE GO KARTS

Pat Best and George Lavigne race side by side at over 200 miles per hour in their rocket-powered go karts.

One night, George called me to ask if I could build him a larger fuel tank to allow him to go faster. He was determined to beat Pat, but I gently reminded him of the deal we had made back when I first built the kart and how concerned I was about the speed of the kart if the tank was able to hold more fuel. Not surprisingly, he didn't like my answer, so he turned around and hired someone else to do it.

It wasn't long before Pat did the same thing, because these two were dead set on beating one another. It had gotten personal. They had a match race one night out at Twin City Speedway, and things got completely out of control as Pat's go kart was clocked at 248 miles per hour through the lights! When he hit the brakes, he got the surprise of his life. Because the small wheels were spinning the rear axle so fast, it started to flex violently until it bent just like a boomerang, causing the rear wheels to completely lock up. The kart made a 360 spin on the other side of the lights before he regained control and brought it to a safe stop. It was at that point that my original concerns became reality, and I realized what a huge mistake they had made by increasing their tank sizes. Here were two men who now had tiny little go karts capable of going as fast as a Top Fuel dragster. "Worried" doesn't begin to describe what I was feeling once that realization sank in. What would happen next?

A couple months later, I was asked to line up a full-blown rocket show for a promoter in Wisconsin, so I booked the *Conklin Comet*, Captain Roller Ball (one last time), and the two rocket go karts at a Kaukauna, Wisconsin, racetrack. George and I drove around town most of the day doing interviews on television, radio, and with the local newspaper to promote the big event. I distinctly recall that there was something different about him that day, and I wasn't sure if something was bothering him or what, but he undoubtedly had something on his mind. I really picked up on it as he was driving, because he was so reckless; I had to caution him to slow down several times. I got busy getting the car ready as soon as we got out to the track, and that's when I learned what had been bothering him so much. He approached me and asked me if I'd be willing to help him run his go kart because he needed someone to pull it back after he did his burnout runs, and his crew member hadn't shown up to help. I told him no problem, that just as soon as we ran the *Conklin Comet*, I'd help him the rest of the night.

George then expressed to me something that I've never, to this day, forgotten. He said to me in a joking way, "NHRA is afraid that if someone crashes in a rocket go kart, the driver's head could come off and splatter blood all over the crowd. That's one of the reasons the NHRA will not approve rocket-powered go karts to run on an NHRA track." He continued with, "This is going to be my last run, Ky, 'cuz when I get home, I'm selling my go kart."

I remember just standing there dumbfounded, not knowing what to say. We made our run with the car and set a new track record. True to my word, I went over to George's trailer to help him fuel up and get on his fire suit. Just before he jumped into his kart, he walked over to Pat Best and said something to him and then came back and climbed aboard. He told me to push him to the starting line so he could start his warm-up burnouts, a standard routine that's done before any of the hydrogen peroxide rocket–powered vehicles run. It slowly lets the fuel flow into the motor, which then heats up the catalyst pack. When you hear the excess fuel start to pop in the motor, you then hold onto the brake and hit the throttle about halfway. You can always tell when a rocket is ready to run by the way it sounds.

What was slightly different about the go kart guys is that they wouldn't hold the brake, so they'd almost leap from the ground when they did their burnouts. George was clearly not focused as he did his burnout. At about 20 feet down the track, he slammed on his brakes and made his kart slide sideways. I ran out and pushed him back to the starting line and cautioned him to be more careful. He did the same thing a number of times until the starter finally lined them both up. The Christmas tree started its countdown to green. George got a head start on Pat, but at about the halfway mark, Pat had almost caught up. Just as things were getting really exciting, the inexplicable happened. George's kart suddenly swerved sharply to the right, almost hitting

SIDE-BY-SIDE GO KARTS

Pat's. As he overcorrected to avoid crashing, he drove sideways and hit the slick grass, which carried him directly into a culvert and then head-on into the steel guardrail. I was standing on the starting line absolutely horrified and actually thought I saw a head go flying down the track. George's earlier words had been etched in my mind, but I was relieved to hear someone yell out that it was his tire that had come off when he hit the guardrail. The rescue crew arrived immediately, and he was alert and talking to them as they rushed him to the hospital. I went to our motor home and dropped to my knees and started to pray. I could not believe what had just happened.

No one, including myself, thought he'd succumb to his injuries, but none of us knew at the time that there was a huge cut to his leg that was hidden under his thick fire suit. To our horror and amazement, George bled to death on the way to the hospital. The news shocked us all and was a devastating blow to everyone, especially Pat. When I spoke to Pat afterward, I asked him what George had said to him before the race. He told me he'd gone over to let Pat know this was going to be his last race and that he wanted to give the fans a thrill by crossing each other halfway down the track. I believe that's what he was trying to do before he lost control.

My two sons were among the thousands of people who witnessed that horrible accident. George was a true thrill seeker, so young, in his 20s and so full of life. He always knew what he wanted and always worked hard to achieve it. Because I knew him so well and what he stood for, I truly believe he was at peace when he passed on. The tragedy changed Pat's life forever, and he ended up selling his go kart to Fred Goeske, who sold it to Rosco McGlashan, who then took it to Australia and set the go kart record there at 252 miles per hour. Afterward, it was donated to an Australian museum.

George Lavigne and Pat Best with their hydrogen peroxide rocket–powered go karts.

Chapter 13

Lee Taylor

While I was at his Torrance, California, shop one day, Craig Breedlove introduced me to Lee Taylor. With a two-way average of 285.21 miles per hour, Lee held the water speed record, which he set in a Hustler J46 Westinghouse-powered jet boat out on Lake Guntersville. I was more than a little impressed by that, but it seemed to me that every time I heard about someone going after that record, they were killed. I have never been interested in boat racing myself, as water is very unpredictable and unforgiving.

I shared my impressions of the sport with him and then asked him what had been the worst crash he had ever been involved in.

Without hesitation, he told the story of a time he was running the Hustler at Lake Mead in Arizona. On his first run, he destroyed the parachute, so he decided to make the second run without one. He was going over 290 miles per hour when he backed off the throttle, but the boat didn't slow down at all; in fact, he said it felt like he was flying through the air. He said the shoreline was fast approaching, so he jumped out of the boat, which was traveling at a super high rate of speed, and into the water. Sure enough, the boat climbed up the embankment and ended up in some shrubs, with surprisingly little structural damage. However, Lee had not been so lucky. He had gone end over end and proceeded to cartwheel over a pile of rocks and right back into the water, where he lost consciousness. He suffered a severe brain concussion, a broken back and numerous other broken bones, and had one eye hanging out of the socket.

I guess everyone there at the scene couldn't believe the horror of it all and scrambled to do what they could. They called the sheriff and worked feverishly to save Lee, but *no* one was prepared for what happened next. Just as they were trying to deal with what they saw before them, things got increasingly worse, as yet another tragedy unfolded before their eyes. As the rescue helicopter attempted to land on the embankment, it crashed and tumbled right over on top of Lee, causing even more severe injuries! They immediately dispatched another chopper, managed to get him on board, and rushed him to a nearby military base. By that point, Lee's condition was extremely life-threatening. Yet as they landed and the emergency team assessed him, they instantaneously told the helicopter pilot they couldn't take him there; he'd have to be transported to another hospital that was better equipped. He was air-lifted once again and finally delivered to a facility where he received the help he so desperately needed.

Lee told me he was in a coma for 18 days, and when he came out of it, the doctor repeatedly asked him two questions: "Where are you from? And what is your name?" Lee said he could barely speak the word "California," and he couldn't remember the answer to the second question. He spent about six months in the hospital recuperating and had to learn how to walk and talk all over again; he never regained his eyesight in the one eye.

After hearing that horrible story, I told him it sounded like he'd used at least 10 of his 9 lives. He shocked me when he said he was going to build a new boat that would go over 400 miles per hour. Boy, and I thought *I* was crazy!

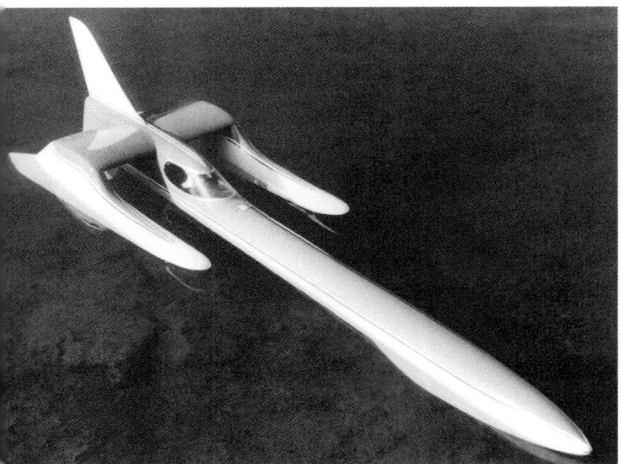

Lee Taylor's test model of the *U.S. Discovery 11* hydrogen peroxide–powered rocket boat.

Lee was at Craig's place that day to look at one of his J79 jet motors left over from his land speed record project. While he was doing that, Craig told him about my rocket cars, and he suddenly became very interested in some of my projects, so I gave him my calling card before I flew back home to Minnesota.

Lo and behold, I heard from Lee a few weeks later, and he offered to fly me back out to California to talk about his new boat project. He wanted me to design and build a rocket engine for it. Once again, I found myself California bound and ended up staying at Lee's home with him and his wife, Dorothy, for a week. On June 12, 1975, I signed a contract with the *U.S. Discovery 11* team to design the propulsion and fuel system. They paid me $3,000 over my cost and awarded me a new Cobalt boat and enough materials to build another rocket for myself. That was a heck of a deal, and I was really excited to get started and wasted no time. In the week I was there, I sketched out a boat design that looked similar to the land speed record car that I was in the process of building—except instead of wheels, I added sponsons with skegs on the bottom so the boat would drive itself up on top of the water.

I spent a considerable amount of time working with Lee on the design and even went so far as to have him sign a confidentiality agreement with me. I did so because, at the time, there were a number of other people building rocket cars, and I didn't want to educate him and then have him start building rocket engines for others. He built a scale model of the *U.S. Discovery 11* and used Estes model rockets to prove the design. With a few minor changes, the model boat seemed very stable. To raise money to build the real thing, he put together a team of boat enthusiasts and promoters, including Bob Kachler, a well-known promoter and marketing consultant, who had put together many sponsorship packages for some of the biggest race teams in the country.

I was really into the project and worked as hard as I could with Lee to make it a success. However, my overwhelming optimism soon turned to disbelief: I learned at one of our meetings that on March 14, 1977, Arthur Williams applied for a patent that in many ways resembled the design I had originally presented to Lee. I immediately felt I had been used, and I lost all interest in the project after that. Lee was really surprised by the change of events and told me he had no choice but to team up with Mr. Williams. I don't know what kind of agreement the two came to, or even if they did for certain, but he told me that over the years he had had many setbacks and frustrations, so he was determined to get this boat built, no matter what. I understood, but it didn't change things, so I bowed out completely, yet followed their progress from time to time.

I learned that Williams was granted the patent June 20, 1978, and that they'd chosen to go with a lighter monocoque design than I had first proposed. When I heard that, I was concerned about the boat's ability to handle the stresses of a high-speed water run, so I once again stepped up to the plate. I proposed to Lee that they build a boat out of chrome-moly tubing with a stressed-aluminum and stainless-steel skin, filled with foam, with an enclosed driver's capsule. He didn't listen. I continued to follow their progress and was rather impressed when I saw that the June 1980 *Popular Mechanics* magazine had a feature article on the *U.S. Discovery 11* boat. However, that all changed on November 12, 1980, when Lee tried for a new water speed record. He was traveling 269.85 miles per hour on Lake Tahoe when the boat disintegrated. He lost his life at 45 years of age while pursuing his dream.

Chapter 14

Human Fly

In 1977, I was contracted to build a rocket-powered motorcycle capable of jumping over 27 buses. The jump was to take place in the Montreal Olympic Stadium, as a half-time show for a concert featuring Gloria Gaynor and a number of other disco stars of the 1970s. The daredevil rider was Rick Rojatt, otherwise known as the "Human Fly." At the time, Evel Kneivel held the record jump of 13 buses, and Rick wanted to beat it badly. Rick's claim to fame at that point in his career was an astonishing stunt he'd performed over the Mojave Desert, where he'd wing-walked on a DC-8 and made two low-flying passes at 250 miles per hour, a nearly impossible feat and truly death-defying. The other very unique thing about this off-the-wall daredevil was that he was never seen out of costume and kept his true identity a secret by wearing a red mask and a white cape, identical to the comic book action hero.

Ky and the Human Fly stand next to the newly delivered rocket-powered Harley.

HUMAN FLY

From the moment I met this guy, I was convinced that misfortune had an APB out on him and was destined to track him down, especially when he told me he wanted to attempt 36 buses. I managed to convince him otherwise when we discussed the fact that in order to do something that remarkable, he'd have to hit the jump ramp at well over 100 miles per hour and continue to burn the rocket a couple more seconds after takeoff.

I explained that it was definitely possible but that acceleration of that magnitude in such a small area would launch him head-first through the concrete pillar at the opposite end of the arena. That conversation resulted in his finally accepting the challenge for 27 instead. He'd still have to travel at 80 miles per hour, and it wasn't that there wouldn't be a crash; I was certain there would be. It was just a matter of how bad it would be by the time he reached our nets and the huge air bag we'd have set up. I knew the decrease in speed and thrust would make a huge difference, and I felt we could pull this one off with minimal damage to the bike or to Rick (hopefully).

The guy was determined, and since my business at the time was working with stunt people, daredevils, and people with death wishes, I remained intrigued and as optimistic as possible, praying I wouldn't fall witness to the hand of death swatting the Human Fly. I became even more concerned when I received a phone call from an insurance broker, Bruce McCaw, who called to thank me for throwing a lot of business his way. When I asked him what he meant by that, he told me he'd just issued a life insurance policy on the Human Fly and that Lee Taylor had been a client of his as well. That really got me thinking.

The Human Fly's rocket-powered Harley with a new seat, gas tank, and paint job.

ROCKETMAN

The Human Fly does a wing-walk on a low-flying DC-8 in the Mojave Desert.

As we prepared for this stunt, it soon became obvious that one of the biggest challenges we faced was the space constraint in the arena; there was no room to accelerate to the speed required, so I came up with a plan. I'd build a rocket-powered motorcycle that would sit right at the bottom of the ramp instead of making the usual fast-and-furious approach. All the Fly would have to do was get on, wave to the crowd, press the button, say a quick prayer, and hang on for dear life!

Rick liked the idea and agreed to try it, so he sent me a brand-new 1977 Harley-Davidson XL-1000 Sportster, a true black beauty, to build from. I put exactly three miles on it, and then the fun began. I yanked out the engine and built two 1,500-pound-thrust hydrogen peroxide rockets, which I mounted side by side, directly underneath the fuel tank. Other than the two polished stainless-steel rocket motors hanging off the back of the bike, I left everything intact, including the headlight and taillight, to make it look completely stock. By the time I finished, this refined machine boasted 6,000 horsepower. If a guy were to take this thing out to the local drag strip, hold the throttle wide open, and hang on hoping the tires didn't fall right off, he'd be capable of going well over 300 miles per hour in the quarter-mile. I let Rick know the bike was ready, so we delivered it to Montreal. We all met up the day before the big event, delivered the equipment, and met with the promoters to discuss the plans.

They had hired contractors to put the jump equipment together for us, which I wasn't really happy about but finally agreed to. We went over the stunt as thoroughly as possible and, much to my amazement, Rick didn't want to do any practice runs at all. He just sat on the bike, admiring it, determined to wait until the time came. I gave him step-by-step, detailed instructions on how to operate the rockets, and he just took it all in, nodding as I went along. I knew he understood what I was saying, but I hardly slept that night because I was always so safety conscious and typically rehearsed stunts many a time before actually performing them. I was uneasy with this particular situation.

Jim Deist, Dar Robinson, and I arrived at the arena bright and early. Much to our surprise, the blueprints for both the jump and receiving ramps were obviously not adhered to, as there were major flaws in both of them. The jump ramp was much too steep, which would cause the rocket bike to come off it at the wrong angle and stall. I was even more concerned about the receiving ramp, though, as the last 10 buses were supposed to be covered by plywood extending to the ramp. What we found instead was a plywood ramp that was about 6 feet above the buses with exposed, steel cross-members. I told the promoters that their contractors, or whoever it was that built these things, obviously didn't follow the blueprints we'd provided and I was not going to fuel up the rocket bike until proper changes were made to the receiving ramp. It turned into a major ordeal by the time we did our last safety inspection, which forced the show time to change quite a bit.

We watched all the prestunt entertainment, but there was no sign of the Human Fly anywhere. As intermission approached, we were all really apprehensive and, to be honest with you, I truly couldn't believe my eyes when Rick and his entourage entered the arena. I had secretly been hoping he'd maybe gotten up that morning, looked in the mirror, and asked himself, "Do I really want to die? Is this really a good idea? Do I really want to break Evel's record this badly?"

That was not to be, though, and as the promoter announced the stunt and Rick took center stage in full costume, the crowd went absolutely wild. I stood in awe as he hopped on the motorcycle, waved to the crowd, looked over at me, gave the thumbs-up, turned on the safety switch, and slowly opened the throttle.

The rocket bike started up the ramp slowly at first, and then the Human Fly pinned the throttle wide open. The cloud of smoke was like a dense fog under a full moon as it engulfed the nearly pitch-dark arena. The superheated steam shot out the back as the bike climbed up the ramp and, instead of launching forward into the air, went much higher than it should have and nearly straight up. Because of the wrong angle, it stalled when he let off the throttle, and the rear end dropped, nearly arching the bike completely backward as it hit the receiving ramp hard, before crashing down on him.

My heart pounded hard against my chest as I stood there, witness to the crash of all crash landings right before my eyes. A hush fell over the crowd. We all feared the worst, certain that nobody could have possibly survived such a brutal landing. Rescue workers ran over to Rick's limp body, and we stood in silence, watching helplessly as they assessed his injuries. Time stood still, and it seemed an eternity before we were given an update by one of the rescuers that he was going to be okay. He'd survived the crash, and he'd done it—he had broken Evel's record, but not without paying the price. He waved to the crowd as he was carried off on a stretcher, suffering a broken ankle and a couple other injuries.

That jump and my rocket bike went down in the history books and then, as time went on, I lost track of both the Human Fly and the bike. I'm happy to say, however, that this story has a rather unique ending. Even though I had long since gotten over owning that fine machine, I received a call not long ago from a stuntman friend of mine, Bubba, a renowned and darn good motorcycle jumper himself, with some astonishing news. He told me he had picked up a *Trading Times* magazine while in Florida and was dumbfounded when he spotted a motorcycle in it that he could have sworn was my original rocket bike. He gave me the phone number, and I immediately contacted John Werner, the owner, who attested and confirmed that it was in fact the bike used in that incredible stunt. I told him to name his price; I wanted it back and *badly*, so he agreed to sell it to me for $6,500.

I sent a good friend down to pick it up and am pleased and proud to say it is now resting back in my rocket shop, where I built it some 23 years ago. They had it rebuilt after the crash, and it looks good as new, in great shape. The only thing they changed from its original design was to add a gas tank. I find myself staring at it frequently and reminiscing about that history-making event quite often. I'm happy to say that if you ever hear of anybody looking for a bike that can jump 27 buses, I've got just the thing. But there's one condition: The rider needs to see a psychiatrist first!

The Human Fly is swatted but survives.

Chapter 15

Kitty O'Neil

Bill Fredericks, Kitty O'Neil, and Hal Nedham standing next to their land speed record car, the *GMI Motivator*.

My years of doing promotions and running the rocket car across the country with Ed Ballinger provided me with a very comfortable living, but it just didn't satisfy my inner drive and desire to turn my ultimate dream into reality: to build a car that would go through the speed of sound. I thought about it night and day, yet every time I was about to seriously put my mind to it, opportunity came knocking on my door, and I found myself heading off in other directions.

Call it fate, or call it luck, but one such opportunity took me by storm, and I would have been a complete fool to pass it up. I became fortunate enough to team up with the one and only Ms. Kitty O'Neil, who at the time held the women's land speed record of 512.710 miles per hour and also the women's water speed record at 275 miles per hour. I don't know that I've ever been as impressed with anyone as I was with her, and I wasn't alone. Kitty O'Neil was truly a national celebrity and, due to the fact that she was both hearing impaired (deaf, actually) and a woman, always received a lot of really good publicity. In fact, she was so legendary and popular that she even had a major sponsorship with Mattel Toys, which produced a series of dolls called "TV's Star Women." They were Barbie-sized dolls of Cheryl Ladd, Kate Jackson, and Kitty O'Neil. Part of that deal wound up trickling down to me when I entered into a contract with Kitty's company granting me a percentage of the Mattel toy deal, as they planned to produce a toy replica of every vehicle I built for her.

I couldn't help but wish that I'd had this opportunity much earlier in my career, because I had sure built a *lot* of vehicles! With the chance of a lifetime before me, I wasted no time and dove right in.

I began by building a 7,500-pound-thrust motor for a new rocket-powered drag car with a 30-gallon H_2O_2 fuel tank that we aptly dubbed the *Mattel Rocket Car*. It seemed like no time at all and we were off. Our true adventures began as we hit the racing circuit fast and furious and traveled from coast to coast all across the United States, astounding the crowds (and even ourselves) by breaking records everywhere we ran.

Along the way, we encountered many, many race teams in pursuit of the same thing we were—to go as fast as was humanly possible and to survive to tell about it. One such racer was Tim Perry, who had built a beautiful Funny Car that was powered by one of Bill Fredericks' rocket engines. Tim was a true competitor, but unfortunately, the car only made a few runs before disaster struck. He was booked into Bandimere Raceway in Colorado, where the track was built on the side of a mountain and was very short to begin with. In fact, when Kitty ran there, she was clocked at 327 miles per hour and nearly ran right off the end of it. One of Tim's crew members had inadvertently forgotten to pull the safety pins out of the parachute packs before the run, so after making his exhibition run down the track, Tim hit the parachute lever and the chutes didn't deploy. Tim crashed his car hard at the end of the track, and the emergency system vented off the hydrogen peroxide all over him. He suffered severe burns to his legs, and the car was totally destroyed. Tim's accident became only the tip of the iceberg for the racing pioneers who were out there pushing themselves to the limits.

It wasn't long afterward, while I was out in California, that I received the very sad news that Jerry Hehn had been killed in his rocket car. I was told he had installed a new catalyst pack and was testing his car in a sand pit in Wisconsin. Apparently, the car wasn't restrained when he hit the throttle, which then stuck wide open, catapulting out of control with no way of stopping it. The car crashed into the side of the sand pit, and the steering wheel pierced Jerry's chest, killing him instantly. I began wondering how there could be so many fatal accidents with rocket cars, while we never even encountered so much as one engine failure; drivers were being killed because of simple mistakes in their judgment.

Fred Goeske later purchased Jerry's car, then took it out to California and rebuilt it. With Fred as owner and George Hadebeck as driver, they made quite a team for a while. In 1980, while at the track in Union Grove, Wisconsin, George also encountered a problem stopping and wound up going off the end of the drag strip, completely destroying the front end, but he managed to walk away unharmed. They say bad luck comes in threes, so while the car was once again rebuilt, it was then retired, robbing fate of its third chance.

Kitty and I decided it was time to go after the *Pollution Packer*'s records and break them. Even though I'd be competing against myself in a way, since I had built and owned the car back in 1972, I felt that records are only made to be broken. Besides, I knew Kitty was driving a better, more powerful car now and that she was more than qualified at handling the driving chores. Our first step toward putting forth the challenge was to really see what our car could do, so we contacted Dave Petrali, the official timer for the USAC and the FIA and asked him to come out and time us at the Mojave Desert in California. He agreed to help us out, and we were psyched.

The day before the official timed runs, Kitty wanted to make some practice runs and check out the lake bed, so, along with Dave, we headed out to El Mirage Desert and invited a number of my Hollywood friends and fellow racers along for the ride. We could clearly see that she had over a 2-mile stopping distance, so I knew if we were going to make an all-out run, this was definitely the time and the place to do it.

ROCKETMAN

The *Mattel Rocket Car* gleams out in the El Mirage Desert.

I set the tank pressure higher than I had ever set it before and told Kitty if she felt like the car was having any kind of problems at all, to shut it off immediately.

On her first run, we encountered trouble with the timing clocks. They recorded her speed, but not her elapsed time, so we refueled and made another run, with the same results. Clearly, not all the lights were working, and Dave and his crew were completely perplexed. After some lengthy discussion and lots of head scratching, we decided to experiment by having a regular car run through. Much to our surprise, the clocks worked just fine, so we figured the problem had somehow fixed itself and got Kitty into position once again. This time, I watched carefully as the crew refueled and saw that the car used 30 gallons of hydrogen peroxide per run, which meant it was making three-second runs. I went over and asked Dave if the clocks were capable of recording a three-second run. He said he didn't know for sure, and I told him I was pretty sure that was the problem. The clocks were old and had originally been made for land speed record stuff. He concurred that they were in fact the same clocks used when timing the *Pollution Packer* five years earlier, and that's when it hit me like a rock. Back then, the *Pollution Packer* wasn't running in the fours, let alone the threes, so we were pretty convinced those clocks weren't going to do us justice but decided to give it one last shot to test our theory.

KITTY O'NEIL

Kitty sets records in the El Mirage Desert.

With a pretty sure guess that one more try wasn't really going to change anything, we couldn't have possibly known how wrong we were. On her last practice run, things did not go smoothly, and everything did indeed change—suddenly and drastically. Kitty nearly lost total control of the car and, at over 400 miles per hour, got so sideways that I could actually read the writing on the side from where I was standing. I honestly to this day don't know how this meager 98-pound, 5-foot-tall woman did it, but from sideways and unstoppable, she somehow managed to regain control and bring it to a safe stop. That was the closest I have ever seen anyone come to rolling a car without actually doing it!

When Jim Deist, Bonneville Butch, and I reached her, she was climbing out, completely unharmed. The first thing she said as she got to her feet was, "That was fun." Well, it may have been fun for *her*, but it was a heart-stopping event for the rest of the crew and me, and we were just thankful as could be that she was still in one piece.

Both of the rear struts holding the rear wheels were badly bent, along with some frame rails. At first glance, I figured our chance at the record attempts was now over, but that ever-present determination of mine soon kicked in. I decided if we burned the midnight oil and put forth lots of hard work, we could most likely make the repairs in time for the next day's official speed runs. Everyone agreed, but we needed a place to work and the equipment to get it done. Fellow racer Doug Kruse stepped up and offered to let us use his shop, so the crew and I loaded up and took the car over. We worked well into the early-morning hours, and while we were obviously unable to restore the car to its original condition, it was darn close.

THE FAST LIFE

An Insightful Conversation with My Friend, Kitty O'Neil

Ky: So, Kitty, what was it like to drive a rocket car?

Kitty: Oh, Ky, it was such a big thrill and so-o-o-o exciting. The view you experience at that speed is absolutely unbelievable. I would love to do it again!

Ky: Can you describe what the g-forces felt like on you?

Kitty: It's exactly like being hit really, really hard and having the wind completely knocked out of you. The weird thing, though, was that it's a sensation you could get used to. I sure did! I got to experience many forms of g-forces, and my favorite was in a fighter jet. Unlike the rocket car, where you feel it on your chest, face, and brain, flying stimulates your legs first and instantly moves to the rest of your entire body.

Ky: Do you think your lack of hearing impaired either of us in any way?

Kitty: Well, I can't speak for you, but I can tell you that for me, not having to block out noise really helped with my concentration and focus. I didn't find myself distracted by all the sound; only by you and all of your hand signals! To be honest, I love my deafness.

Ky: Can you tell me what you actually thought about while you were driving?

Kitty: W-h-e-e-e-e-e! Wowie!! Oh Boy!!!! Those are just some of the recollections of what I found myself thinking, but before every run, I would pray for God's guidance to allow me to experience a safe and thrilling drive.

Ky: So, you just know I have to ask . . . What do you think of me as a rocket car designer, builder, and friend? And be nice, because this is going in the book!

Kitty: As a friend, I've always thought you were crazy in the world. You always made me laugh, and you were loyal. As an inventor, you were smart; I trusted you 100 percent and had nothing but true faith in you at all times.

Ky: So what has it been like for you now that you've been retired from the fast life for so long?

Kitty: Well, retiring was not easy, but I did it. I'm no longer into exercising the way I used to be. Back when I drove the rocket car, I ran four miles a day to help expand my lungs and condition them to handle the g-forces. A driver is an athlete and must keep in good physical condition, which I did. I'm not saying I'm out of shape *now*, but I did find myself taking life a little bit easier than I used to.

Ky: Thanks, Kitty.

So there you have it! Right from Ms. O'Neil herself.

ROCKETMAN

As we neared completion, we made a unanimous decision to slow the car down in the name of safety and to only attempt to break the original *Pollution Packer*'s records set in 1972 for the quarter-mile and 500 meter. It had proven much too risky for Kitty to go after the kilometer record, as we had clearly seen that the car was capable of going much faster than it was originally designed to. Even though she was adamant about going all out, I knew we were also taking a big chance with the tires on the car, not to mention the problem with the timing clock. Without knowing whether or not Dave had been able to fix it, I secretly set the tank pressure down to 500 psi, controlling the maximum speed, totally unbeknownst to Kitty.

We arrived on location tired but determined and in very good spirits. We were going for the quarter-mile record first, and Kitty needed to make two runs in opposite directions. Both the car and Kitty put on a good performance under the conditions. She made her east run first and, at the end of the quarter-mile, achieved a 5.249-second ET with an average speed of 171.461 miles per hour. We got her all lined up and refueled for her second run, and she took off westward in perfect form. That run was clocked in with a 4.809 ET at 187 miles per hour. Her official average ET that day was 5.029, and her official average speed was 178.962 miles per hour, easily breaking Dave Anderson and the *Pollution Packer*'s prior record of 158.8 miles per hour in 5.666 seconds set on September 30, 1972. Those speeds may sound slow,  but remember that we'd only run at about 50 percent of the car's full power.

Next up were the 500-meter runs, again in both directions. Kitty's west run recorded her at 5.424 seconds ET, with an average speed of 206.207 miles per hour, and to the east she clocked in at 5.343 ET, with an average speed of 209.333 miles per hour, making her official average

KITTY O'NEIL

Ky and Stockard Channing help Kitty put on her fire suit.

ET 5.384 seconds and her official miles per hour 207.739. She once again broke Dave's previous record of 173.9 miles per hour in 6.31 seconds, as well as Paula Murphy's Bonneville Salt Flats record set on October 1, 1972. Amazingly enough, she even beat Vern Anderson's record of 5.492 ET, 203.536 miles per hour set on September 19, 1974, in the *Pollution Packer Bonneville Dragster*. It was an absolutely incredible experience, and we were all ecstatic. This car-and-driver combination had truly been the very best a guy could hope for, and yet another of my many dreams had undeniably come true. It was a proud, proud day in history as our records for National Rocket and International Category C were submitted to the FIA in Paris, France.

Having accomplished what we'd set out to do, Kitty and I both realized we should probably quit while we were ahead and go out on top, so we parted ways. After all, my original dream of building a car that would break the speed of sound was still very much alive in my heart, and I could hardly wait to get started. I headed back home to Minneapolis and spent several very happy months designing and building the frame, but in order to truly finish, I knew I'd have to head back out to the West Coast.

Kitty and I stayed in touch, and she offered to let me ship my frame out to her shop to have an all-aluminum body built for it, but there was a catch. She'd help me if I helped her. Knowing her need for speed, I wasn't at all surprised when she asked me if I could build a rocket-powered drag boat for

Ky Michaelson takes a minute to pose with friends next to his land speed record car at a show.

her from a pickle fork hull that Sanger Boats had donated to us. I figured why not, since I would be there for a while anyway, so once again, I got my bags packed, had my frame shipped, and headed to the airport.

Upon seeing the hull, I ultimately made a number of changes to it, including the addition of a vertical stabilizer on the back. By the time it was done, the thing looked fast just sitting on the trailer. Oddly enough, though, after all the time, money, and work invested, Kitty changed her mind, and we never even ran it. As I look back, there were probably a number of good reasons for not doing so, and I really have no regrets about it and doubt that she does either. As I've said before, I don't trust water—there's no margin for error. If things aren't just right, you have but one guarantee: you *will* have a bad day.

So, with my LSR car in the works and the boat project shelved, you'd think I'd just relax and enjoy some down time, but that wasn't to be. Kitty was as go, go, go as I was, and she wanted to get back out on the circuit, so she challenged me to see if we could make our *Mattel Rocket Car* the *world's* fastest dragster. Without much arm twisting on her part, she got me all fired up, and the wheels started turning.

We knew the only way to become the world's fastest was to head back out to El Mirage Desert and go for more than just breaking the original *Pollution Packer* records. This time, we had to go for an all-out quarter-mile ET and mile-per-hour record. I still felt the car could definitely do it, so we lined up the same production crew we'd had previously and were assured they'd have new timing equipment this go-round. Within a week, we were all set, and it felt really good getting back out there with Kitty. Everyone involved was more than just a little excited, because they all remembered what this car and driver could do, and we were certain we'd pull off the challenge without a hitch.

KITTY O'NEIL

Since the car hadn't been run since our last outing and the new clocks hadn't been tested yet, I made sure to set the tank pressure exactly the same as I had before. As we pushed her into position, I leaned over and begged her to be really, really careful. She smiled at me and just said, "Okay."

She blew down the line and ran a record-smashing 396 miles per hour with an ET of 4.12 seconds. The clocks were definitely working, and things were looking really, really good. When we got down to Kitty, the first thing she said was, "Hey, I want to go *faster*!" She told me she shut the rocket off early just to be safe, which built up my confidence in her. Since I knew she'd listened and was being safety conscious, I boosted the tank pressure to over 800 psi. Kitty made four more runs, all of them with ETs under four seconds, with a best time of 3.22 and a speed of 412.68 miles per hour. Those incredible times indisputably proved our car and driver were truly the world's fastest and that when we set our minds to doing something together, we were just plain unbeatable.

I was staying in a house not too far from Kitty's place and will never forget the day she came pounding on my door, so excited she was nearly breathless. She rushed in waving a piece of paper at me and just grinning from ear to ear, as she held it out for me to see. It was a letter from a production company called Channing Debin Locke that said they wanted to produce a prime-time television special called "Silent Victory," based on Kitty's life story, with Stockard Channing cast as Kitty. She was beyond ecstatic, and I was so incredibly happy for her. She'd suffered a high fever as a baby, which had caused her deafness, yet it had never been an obstacle for her. She had endured many hardships throughout her life and through it all had still accomplished more than any other woman I have ever had the honor of knowing. The title for the show captured everything she stood for, and she was absolutely thrilled at the thought of it. Without hesitation, she happily accepted the deal and couldn't wait to get started.

We soon learned that in order for Stockard to realistically portray Kitty, the production company wanted her to see Kitty in action behind the wheel of a rocket car. She needed to learn what made Kitty tick, mimic her mannerisms, watch and learn as Kitty read lips—all of it. She had a *lot* to learn but had one heck of a teacher, no doubt. We had already accepted a prior offer to perform on a national TV show being filmed out in the Mojave Desert called "Super Stunt 2," which was hosted by Rock Hudson, so Kitty let the producers know she'd be able to show Stockard just about everything she needed to see if they'd come out for the shoot.

It was a thrill meeting Stockard, and I could tell she was really honored to meet Kitty. Stockard commented to me that she'd read a lot of newspaper articles about Kitty and was totally impressed that this meek and dainty woman could be so completely fearless. I told her that was actually just about the only thing wrong with her; in the stunt business, it is not good to be without fear, but throughout all the years I'd known Kitty, I knew *that* wasn't going to change anytime soon.

Rock Hudson and Ky check out the rocket-powered Corvette before its record-breaking run.

Stockard watched with great intensity as Kitty prepared to drive a 1977 Corvette with a 7,500-pound-thrust rocket motor in it. I could see she was studying every move and was nearly speechless as she witnessed Kitty set two more world records during the filming of the show. Kitty ran 365.21 miles per hour in 3.58 seconds, which was the first 300-plus miles per hour run *ever* made in a Funny Car and the first Funny Car to run under 5 seconds.

The production crew got much more footage than they'd bargained for that day, and Stockard had good reason for her speechlessness. For some unknown reason, Kitty didn't shut the rocket motor off before the lights, which were the instructions I had given her, and in fact drove the car considerably past them. I'd say the car had to be going over 400 miles per hour before she hit the rough part of the desert, horribly out of control and unable to stop. We all looked on helplessly and watched in horror as she rolled and then flipped end over end until she came to a car-crushing stop. With the ambulance sirens blaring toward the scene, my feet didn't touch the ground as I ran to my truck and drove full speed ahead to the end of the course. I just kept thinking over and over that Kitty was dead; I was gonna find her dead, and what would I do? My dearest friend and I'd lost her, I was sure.

As I screeched to a halt at the site of the crash, I was in a complete panic and couldn't even think straight, but I knew I had to get to her and *now*. I saw the paramedics had reached her as I ran to the car and, in a moment of silence that I'll never forget, dropped my jaw as my dear friend came crawling out of the roll cage under her own power and seemingly in one piece. I'm certain my eyes were popped out of my head. I tried to speak, but she beat me to it, and I'll *never* forget the first words out of that woman's mouth.

She said, "See, Ky? I *told* you I know how to roll a car over. Oh, and don't worry about it . . . I'll pay for the car."

Oh man, but I was relieved, and I just embraced her and thanked the good Lord over and over again. It was a moment I know I'll never forget and still amazes me to this day. There she was, not a scratch on her, crawling from a car that was nothing more than a heap of wreckage, completely destroyed. Needless to say, while the crew and I were completely ecstatic that she was okay, we were *not* too happy about her exploits, and after she had a chance to get her bearings straight, we gently let her know so. I have to say, too, that Stockard was totally in awe at what she'd just seen and had no trouble capturing the spirit of Kitty after that. Both the stunt show and the movie were huge successes, and Stockard's portrayal was a stellar performance.

Kitty on her way to setting a new speed and ET record in her rocket-powered Corvette at the El Mirage Desert track.

And just for the record, in case you're wondering, Kitty never *did* pay me for the car. Twenty-five years later, it's still sitting in a garage in Minneapolis in a big tangled mess, waiting to get fixed. Amazingly, too, all those years later, I received an e-mail from Sam Thompson, the paramedic and ambulance driver who'd been on the scene. In it, he wrote, "I got to wondering if you had ever heard what Kitty said to me as I got to her before the cameras, etc., got there. As I got the all clear from her, she said, 'Get Ky. Fix Car. FUN!!! Do again.'" That's Kitty for ya.

I continued to work on my LSR car out in her shop but found myself running low on money, so I made a decision to sell my *Mattel Rocket Car*. Dave Henderson and Ed Ballinger wanted it, so they agreed to meet me at the Rocket and Jet Car Nationals in Union Grove, Wisconsin, to finalize the deal. I told them I'd deliver the car to them there under one condition. Nearly

Ky stands next to his land speed record car.

every rocket car still running would be at the event, and I wanted to see it run *one* last time as mine. They gladly accepted my deal, so, as they say, I was off to the races. I said my goodbyes to Kitty and left my LSR car in her shop, along with our boat, and headed up to Minnesota to pack up the rocket car and drive to Great Lakes Dragaway. Fred Goeske, Sammy Miller, John Allen Hudson, and Brad Proffitt were all there, and I got a huge kick out of just reading the sides of everyone's trailers, because they all said the same thing: "The World's Fastest." I knew better and, while I still owned the bragging rights, was determined to prove once and for all whose car truly was the fastest in its time.

Now Ed had never driven the car with the 7,500-pound-thrust motor in it, so I proceeded to tell him that this was going to be the fastest and quickest run he'd ever made and that he should definitely shut the motor off at least 250 to 300 feet before the lights. He was on pins and needles as he got ready to run and could hardly wait. Thankfully, he followed my instructions and, on his first run, shut the rocket off early and coasted through the lights at 347 miles per hour with an ET of 4.48 seconds. He told me later that he couldn't believe the speed and that when he'd shut off the motor, it was just like he'd slammed on the brakes. His run was the fastest quarter-mile mile run ever made at an NHRA-sanctioned event, once again breaking the *Pollution Packer* record of 344.820 miles per hour in 4.62 seconds. Although NHRA posted a maximum speed limit of 330 miles per hour on the rocket dragster, there were occasions when a driver accidentally went over, and this was one of those times. For first offenses, drivers were given a warning, but if they continued to break the speed rule, their license would be suspended.

Speaking of speed rules, one of the funniest excuses I had ever heard for breaking them came from Fred Goeske while he was driving the *Texas Starship*, which I had originally built for John Allen Hudson at Orange County Raceway. It was also powered by one of my 5,000-pound-thrust rocket motors, and Allen had run it for about three years, reaching his all-time best of 328.76 miles per hour in 4.80 seconds in October 1978. On *this* day, with Fred in the cockpit, it went 353 miles per hour in 4.55 seconds. Needless to say, Bernie Partridge, the NHRA rocket car program director, was not too happy. But Fred was known for his off-the-wall humor, so he proceeded to tell Bernie the light on the dash was burned out and he couldn't see the speedometer. Unbelievably, Bernie actually fell for it, even though there *was* no light on the dash and there was no speedometer. Fred's quick thinking helped him keep his license, and I still laugh when I recall that little tale.

Once I sold the Mattel car, I got sidetracked on many other rocket-building projects, and before I knew it, it'd been about two years since Kitty and I had spoken, as we'd completely lost touch with each other. After calling all over the country, I finally tracked her down in Texas. She flew up to Minnesota for a visit and asked me where the rocket boat and land speed car were. I told her that to the best of my knowledge, they were both still out in her shop in Fillmore, California. That didn't sit well, as she wasn't so sure, and she said, "Let's get out to California and bring them back here for safekeeping." We booked a flight and headed out. Much to our surprise, they were both gone! We immediately reported the theft to the Fillmore Police Department, and to this day, neither the car nor the boat has ever been returned to us.

Ed Ballinger sets a new national record of 347 miles per hour with an ET of 4.48 seconds at Great Lakes Dragaway in Union Grove, Wisconsin.

Chapter 16

Extreme Rocket-Powered Vehicles

In the days before the rocket era, speed freaks tweaked and modified existing engines and vehicles in every way imaginable to go faster—and maybe to make a personal statement too. The rocket engine pushed those speeds to a new level but didn't change the ingenuity or personalities of speed's apostles.

Throttle junkies from every background found themselves in possession or at the wheel of rocket creations during their heyday. They set records and expanded racing lore with extraordinary moments, both jubilant and heartbreaking. Only when the supply of hydrogen peroxide began to dry up for rocket racers in the early 1980s did this new breed of land pilots disband onto the pages of racing history.

EXTREME ROCKET-POWERED VEHICLES

Raul Cabrera and Ron Poole

I met Raul Cabrera and his wife, Maxine, in April 1976 at the Stanford Rotary Club Air Show in Stanford, Florida. They were at the show to watch us put on a couple of exhibition runs with the *Conklin Comet*. Raul told me he and his partner, Ron Poole, were building a rocket-powered Volkswagen for exhibition racing at drag strips on the East Coast. Of course, the second anyone talks to me about rockets, I'm all ears. Raul told me they were powering their car using solid-propellant motors, much like the propellant used in the boosters of the space shuttle. I immediately expressed my concerns with that source, asking how they planned to shut the engine off in an emergency situation. We discussed the fact that, just like the space shuttle boosters, you cannot do so until the fuel actually runs out, and he agreed that posed some risks. I then explained to him that solid fuel's lack of shut-off capability had been my reason for choosing hydrogen peroxide to power my cars. I reflected back to the first time I had heard of someone using solid propellant. It was during a visit I had made to the public library, where I read an article about Fritz Von Opel, who drove the very first rocket-powered car back in 1928. He used signal rockets, typically used for ship distress, mounted in a cluster behind the car.

As it turned out, I never actually got to see the creation Raul had so vividly described to me. I don't know why or how, but those conversations wound up being our last, as we lost contact with one another for many, many years. I didn't hear from Raul again until July 2003, when he sent me an e-mail. After talking to him once again, I felt compelled to include the story of his rocket-powered Volkswagen in this book because it's so interesting and really demonstrates how rocketry has evolved throughout the years.

Raul became interested in rocket propulsion during the late 1950s, when amateur rocketry gained its popularity, influenced by those early years of the space program. He was building and launching zinc and sulfur rockets as a hobby, but his direction for the use of rocket power didn't become clear until 1966, when he became acquainted with Ron L. Poole. Together, they started casting rocket motor propellants using polyester polymer (boat resin) mixed with potassium chlorate. Ron attended Auburn University and studied in the field of chemistry, researching polymer science, which led them to develop their own solid-propellant rocket motors, each contributing to its eventual success. They began making aerial rockets, moved on to rockets on skateboards, and then on to a go kart model. They dubbed their work "Project AZIDE."

The *Frightnin' Lightnin'* rocket-powered Volkswagen.

After they developed what seemed to be a very successful solid propellant, their first rocket-powered car made it to the drawing board. The team decided to build a car using a Volkswagen body on a tube frame and mounted a diamond-shaped cluster of 4-inch rocket motors to it, each using 30 pounds of solid propellant cast with a tubular grain design. They called their creation *Frightnin' Lightnin'*.

On March 31, 1975, with a number of their good friends looking on, Raul and Ron had the car ready to make its debut and perform its first test run. They positioned themselves on an old abandoned highway and, with Ron behind the wheel and Raul in the passenger seat, Ron pushed the ignition button. In that split second, all hell broke loose. The top motor detonated, completely disintegrating the entire body of the VW. Concussions knocked them both unconscious, and they were rushed to Lakeland Hospital, where they spent two weeks recuperating.

Once they fully recovered, Raul and Ron began their investigation into the cause of the accident. Upon further examination of the unburned propellant, they concluded that not only had the polymer fuel become hard and brittle during storage, causing cracks in the propellant while under operating pressure, but the oxidizer used was also the wrong type, which added to the detonation that led to the car's demise.

The team decided to cool it for a while, and it was during their hiatus that Raul took the opportunity to settle down and marry Maxine. A year after his wedding, Raul contacted Ron to see if he was interested in trying to overcome the propellant problem and build another car. He agreed, so they set things in forward motion once again and, after about six months of ordering different samples of raw materials and testing them, they formulated a propellant using 200-micron rotary-rounded ammonium perchlorate (AP) oxidizer and 15 percent zinc powder, which reduced the acidity of the AP. They were sure this new formula would allow them to perform in the presence of a crowd of people. The fuel, called Paraplex P13, was a super polyester/styrene monomer, with 200 percent flexibility and 6,500 psi strength. After a series of tests, they never experienced another motor explosion and, with a five-spoke wheel propellant casting mandrel, designed by Raul, they gained enough confidence to actually begin building another car.

Raul and Ron didn't want to reinvent the wheel, so to speak, so they started out with a standard Volkswagen, which they stripped down. They then added a roll cage and mounted what looked like a military rocket right through the middle of the car, complete with fins sticking out the back. They proudly named this car the *Vulcan Shuttle*. After three years of hard work, they made their debut run with the car in September 1978 at Hernando County Airport in Brooksville, Florida.

Ron drove the car most of the time at a number of tracks around the southern states, but Raul took on the driving chores at Capital Raceway in D.C. It was a night run, and the track lights were off, adding to the difficulty of driving the thing. As soon as the motor shut down, the car's front end suddenly became completely airborne. The *Vulcan Shuttle* was going 188 miles per hour when Raul hit the parachute, then the brakes, and the car slipped sideways, crashing hard to the ground. Raul later said the force had been so intense, he thought he'd popped his eyeballs right out of his head. They'd stayed intact, but they were blood red and really sore for about two weeks afterward. The shoulder straps had broken his collarbone, but other than that, he walked away from the crash and considered himself lucky.

The team picked up the pieces and rebuilt the car one more time. Because of Raul's last tumultuous ride, Ron was put back into the driver's seat once again. He had the most experience, and Raul had pretty much lost interest in driving altogether. Ron made approximately 80 more successful runs before disaster struck. On March 4, 1981, they took the solid-fuel rocket VW to Vaudeville Airport in Tampa, Florida, to test it. In a horrifying instant, Ron lost his life as the thrust of the rocket engine pushed down on the front end, breaking a tie rod, and sending the car end over end with so much force that Ron was ejected and died upon impact. As had happened to so many speed pioneers before them, Raul and Ron's legendary efforts—as the first team ever to build a solid-fuel rocket-powered car in the United States—ended tragically.

Bobby Tatroe and Walt Arfons

The "most powerful" rocket car award has to go to Walt Arfons of Akron, Ohio—master builder of the *Wingfoot Express*—and 28-year-old Bobby

Ron Poole thunders down the track in his land-locked missile.

Tatroe, its extremely talented driver. This incredibly crafted work of art was powered by jet-assisted takeoff (JATO) rockets. The Goodyear tire company even got on board, showing its support by designing and building special 35-inch tires capable of handling speeds in excess of 750 miles per hour. It was truly a land-locked missile, disguised as a rocket car, and raised many an eyebrow when it showed up at the Bonneville Salt Flats. Using the force of 15 1,000-pound-thrust JATO rockets, the team made a number of runs at

Bonneville but was unable to break Craig Breedlove's record of 536.71 miles per hour. Sheer determination drove them to enhance the car by strapping an additional five rockets to each side. It then reached 438.65 miles per hour through a measured mile but not before sustaining significant damage. Two of the rockets fell off—one when the car entered the mile and the other elsewhere during the run—and the retrofire system malfunctioned, causing some of the other rockets to fire in both directions. That neutralized the thrust and burned through the aluminum body and electrical wiring. Bobby said at one point during the run, the air speed indicator read 480 miles per hour. At the time of this memorable run, JATO rockets cost $1,000 apiece, so the hair-raising experience of propelling that magnificent car through the measured mile cost a cool $25,000.

Walt has also been credited for building one of the most interesting and potentially fastest drag racers ever built. Because it ran on superheated water, just like we drink every day, it was relatively cheap to drive in comparison to anything else out there. It was a project that took him just six short months to build in his shop in Akron at a total cost of just $7,000. It had a remarkable power-to-weight ratio. The dry weight of the car was 825 pounds, so when you added Bobby, the driver, at roughly 200 pounds and the water for propellant at 600 pounds, the overall weight as it rolled up to the starting line was 1,625 pounds. What was so remarkable about its design was that Walt calculated the 600 pounds of water weight would turn to steam in mere seconds as it sped down the track, and he knew that a steam rocket has an unbelievable amount of instant power, producing almost 5,000 pounds of thrust. He'd heat the water to 400 degrees Fahrenheit, and at that temperature the tank pressure would reach about 300 psi.

He brought the car out to the Akron Municipal Airport for its initial test run, where Bobby soon learned he had his hands full due to the unbelievable g-forces. The car was estimated to be traveling at over 250 miles per hour near the 1,000-foot mark, when a sudden 30-mile-per-hour crosswind unexpectedly ended the high-speed run with a spectacular crash. Thankfully, Bobby walked away with only a few cuts and bruises and a *lot* of respect for the mighty steam-powered car. They decided to rebuild the car, but the second time around they used aluminum instead of the heavier fiberglass. The new car was renamed the *Neptune* and, without ever unleashing its full potential, was driven only once by Duane Landon. Word has it the car was sold a couple times after that and never run since.

Walt Arfons and Bob Tatroe standing next to the *Wingfoot Express* rocket car.

The car was rebuilt, and the rear fenders were changed to hold the parachutes inside.

EXTREME ROCKET-POWERED VEHICLES

Bob Tatroe drives his steam-powered car at the airport in Akron, Ohio.

Larry "Rocket" Welch

Larry "Rocket" Welch, a long-time motorcycle drag racer, switched over to rocket-powered motorcycles in 1975 when he rode the *Heavy Trip*, powered by two rocket motors and built by Arvil Porter. His advertising brochure listed a top speed of 198 in a mere 7 seconds! I saw the bike run once when I was out on the East Coast, and the thing I remember most is the absolutely ear-piercing sound of the motor. I happened to be standing right by the starting line, and the noise of the engine actually hurt my ears for a couple of days. Sadly, his cycle only ran a very short period of time before the American Motorcycle Drag Racing Association and NHRA revoked his license for what they said was "unsafe operation of a vehicle on the drag strip," and also cited him for operating a vehicle of an unapproved type with willful disregard of the rules and safety of himself and others. His rocket cycle career came to a screeching halt when those citations were filed along with several other infractions and violations.

Walt Arfons examines the steam rocket as Bob Tatroe looks on.

Chapter 17

Legends of Their Time

While a lot of people tapped into rocket power and did some extraordinary things, a few rose above the fray with a little something extra in speed, personality, or both. Some of these personalities are covered in other chapters. Others that captured headlines and records are described here.

Larry Welch poses next to his beloved rocket-powered motorcycle.

LEGENDS OF THEIR TIME

Craig Breedlove's rocket car.

Craig Breedlove

Craig has been my all-American hero since the 1960s. He's a man's man and an innovator whose reputation and accomplishments are nothing short of amazing. In addition to setting the land speed record a number of times, he also holds the honor of driving what has to be the scariest rocket car of all time, the *English Leather Special*, a.k.a. the *Yellow Screaming Zonker*. Its bipropellant rocket engine ran on unsymmetrical dimethylhydrazine (UDMH) for fuel and nitrogen tetroxide (NTO) for an oxidizer, fed to the engine with pressurized nitrogen gas. The propellants were hypergolic, meaning they'd ignite instantly upon contact with each other, and just breathing the fumes could potentially harm or kill you. Nevertheless, Craig had the intestinal fortitude to drive that bomb on wheels down the quarter-mile at breakneck speeds, knowing full well that if he crashed, he'd be taken from this Earth. He ate up drag strips in St. Louis, Dallas, and Orange County, California, setting the AHRA record for the quarter-mile, traveling 266.28 miles per hour in 5.550 seconds. That much success prompted him to also go after the *Pollution Packer* records, so he headed out to Bonneville, where he recorded a speed of 377 miles per hour in 4.65 seconds but lost control of the car. In nothing short of a miracle, while the monocoque chassis underwent severe damage that day, Craig was completely unharmed.

ROCKETMAN

Lew Arrington drives his *Captain America* rocket car at the Union Grove, Wisconsin, dragway.

Lew Arrington

Lew Arrington, otherwise known as "Captain America," started his drag racing career in the stock car class, then moved to a gasser, and from there moved on to fuel Funny Cars. He was best known for his series of *Brutus* Funny Cars, and in 1972, *Car Craft* magazine ranked Arrington one of the top 25 Funny Car drivers. He took about a three-year break from racing, but much to the delight of his fans everywhere, he came roaring back in 1976 with his *Captain America* rocket-powered Mustang Funny Car. This low-slung, landlocked missile was the first of a number of Funny Cars to be approved by the NHRA Rocket Car Committee, which granted Lew one of the first licenses to compete in single exhibition runs with the stipulation that there be a maximum speed of 250 miles per hour.

Dick Keller had built the propulsion system for the car, and on May 16, 1976, Arrington premiered in what would become a new era in drag racing at the Madison Township Raceway Park in Englishtown, New Jersey. There, he drove his Mustang to a speed of 241.93 miles per hour in 6.37 seconds, adhering to the 250-mile-per-hour ceiling adopted by the NHRA in the interest of safety. Lew believed in the speed limit and always abided by the rules in all of his exhibition runs.

Dave sits in his favorite ride on the return road.

Dave Anderson

Dave Anderson was a soft-spoken, kind, and gentle family man who lost his life doing what he loved, and that was racing. I had the honor of knowing Dave since our high school days together, where he won awards for his all-around athletic abilities. He was always a competitive person and gave 100 percent of himself to whatever he did. We were both interested in cars and racing. He drove a Top Fuel car, and I drove a top gas car on the NHRA Division 5 circuit for a number of years, until I got him hooked on the rocket cars and he joined our *Pollution Packer* team in 1972. In a mere 18-month period, he made over 80 runs in rocket-powered cars. He established 13 international and national speed records at the Bonneville Salt Flats in Utah, and 12 state speed and elapsed-time records at drag strips throughout the nation.

Dave was the first driver ever to drive a rocket car at an NHRA-sanctioned event and was also the first to run over 300 miles per hour and run in the 4-second bracket. He thrilled tens of thousands of spectators at the 1973 Winternationals in Pomona, California, setting a new track record of 5.35 ET with a speed of 297.02 miles per hour. Dave went on to the NHRA Summernationals and set the record at 4.62 ET with a speed of 344.820 miles per hour. He was very proud of his achievements, and so was the rest of his team. The tragic crash that took his life, along with two other young racers, in Charlotte, North Carolina, was a shock to all. Everyone who ever had the pleasure of meeting Dave was touched by his love of life and his kindness. The racing world lost a real treasure when we lost Dave Anderson.

Vern Anderson

Vern Anderson is another racer who jumped from driving one of those flame-throwing, oil-spewing Top Fuel cars into the driver's seat of a rocket-powered dragster. Vern, no relation to Dave, also drove the *Pollution Packer Bonneville Dragster* for a short period of time, beginning in Rockford, Illinois, where he'd previously broken his back while driving a Top Fuel car in the 1960s. He ran there at 5.447 ET with a speed of 259 miles per hour. Vern ran his first 300-mile-per-hour run at Albuquerque, New Mexico, at 322.58 miles per hour in 5.110 seconds. From there, he went on to Phoenix and ran 4.72 ET at 335 miles per hour. He also set the lowest ET of 4.663 seconds at the Jet and Rocket Car Nationals in 1974. The team brought the car out to Bonneville on September 19, 1974, and broke Dave Anderson's world and U.S. records for the two-way quarter-mile average speed and Dave's records for the world and U.S. 500-meter average speeds. Vern ended up setting eight world and national elapsed time and speed records before leaving the Pollution Packer Rocket Team, when he switched driver's seats with Jerry Hehn, owner of the *American Dream*. That car was powered by a 4,500-pound-thrust Ky Michaelson Space Age Racing motor. He ran 315 miles per hour in 4.90 at the AHRA Grand Nationals in St Louis, beating Craig Breedlove's long-standing record of 265 miles per hour.

Ed Ballinger

Ed Ballinger, dubbed "Captain Eddy" by his close friends, was a natural-born thrill seeker who drove a number of rocket cars throughout the years. He skydived at the young age of 15, made his first solo flight when he was 16, fought in both the Korean and Vietnam wars, crashed and rolled cars in Lucky Cook's Daredevil Auto Show, and hunted

Ed Ballinger stands next to Ky Michaelson's *Freedom Flame* rocket car.

LEGENDS OF THEIR TIME

200-pound wild boars with a wooden spear. Captain Eddy drove the *Freedom Flame*, *Conklin Comet*, *Mattel Rocket Car*, and the *Chicago Patrol* rocket Funny Car. He set many track and state records, including breaking the *Pollution Packer* national record with a speed of 347 miles per hour at Great Lakes Dragaway Rocket and Jet Nationals in Union Grove in the fall of 1978.

Doug Brown's *Flasher* rocket car.

Doug Brown

Doug Brown drove a small sports car called the *Speed World* hydrogen peroxide rocket car, powered by a 1,500-pound-thrust rocket motor that I built. He sold the car, and it was sold a few more times before disappearing from the public eye.

Jerry Hehn

Jerry Hehn's company, Hehn Engineering, built an 850-pound rocket car called the *American Dream* for which I designed a 4,500-pound-thrust propulsion system. Jerry drove the car for only a short period, during which time he managed to set the following track records:

Jerry Hehn on the starting line.

Track location	ET	Speed
West Salem, Ohio	5.15	278 mph
Epping, New Hampshire	4.98	309 mph
St. Louis, Missouri	4.90	315 mph
Englishtown, New Jersey	4.97	306 mph
Indianapolis, Indiana	4.93	258 mph
Orange County, California	4.82	325 mph
Yakima, Washington	5.02	312 mph
York, Pennsylvania	4.96	296 mph
Puyallup, Washington	4.98	272 mph

Jerry then teamed up with Vern Anderson, who handled primary driving duties while Jerry did all the maintenance. The *American Dream* continued to be one of the quickest and fastest rocket cars on the circuit, until the team's quest for more speed came to a tragic end, when Jerry was fatally injured while testing a new catalyst pack at a sand pit in Wisconsin.

Brent Fanning

Brent Fanning and his wife, Vickie, ran a 1977 Corvette rocket car called *Outer Limit*. The old *X-1* rocket motor chamber powered the car, and I built the rest of the propulsion system for it. The Fannings bought the car from the original owner then cut the top off and made it into a roadster. Because there were so many cars with much larger motors, they'd decided they wanted something a little different. They ran the car for a short period of time, until Brent decided to concentrate more on show business than speed. He made

Brent Fanning's *Outer Limit* rocket-powered Corvette puts on a show.

LEGENDS OF THEIR TIME

Brent Fanning's concept rocket car lights up the sky.

radical changes to the body, added flashing lights and a gasoline-fueled flame thrower, and changed the name of the car to *Concept 1*. It ran ETs between 5.20 and 5.50 seconds. The Fannings owned a dairy farm in Texas, so their racing operation was aptly named "Udder Nonsense Racing." The car and its team were hugely popular with the fans everywhere they went and enjoyed their time on the circuit from 1978 to the early 1980s, when the hydrogen peroxide supply dried up.

Sammy Miller

Sammy Miller entered the rocket car racing scene shortly after Dave Anderson's fatal accident in Charlotte, North Carolina, and drove the *Pollution Packer Bonneville Boss* for a short period of time. Sammy set the Seattle International Raceway record on May 31, 1975, with an ET of 4.55 seconds and a speed of 307 miles per hour. He later bought the rocket engine out of the *Pollution Packer* that had sustained the crash and swapped it for the reciprocating engine in his Ford Mustang Funny Car, which he renamed the *Spirit of '76*. Sammy had raced Funny Cars for about 12

Sammy Miller's *Vanishing Point* car during burnout.

years and never saw a profit. Oddly enough, after the swap, he did see a profit, even though the largest expense of running a rocket car is the 90 percent hydrogen peroxide fuel. At roughly $12 per gallon back then and an average of 20 gallons per run, it was a huge expense. Still, his first year with the *Spirit of '76* actually brought in some profit. Maintenance on the rocket cars was very minimal, and Sammy became a very optimistic racer. It didn't take him long to set the rocket Funny Car record at 250 miles per hour in 5.62 seconds. He later sold the car to Fred Goeske, who renamed it the *Chicago Patrol*. Sammy continued to race over in England for a number of years, until he passed away in an oil field accident in 2002.

John Paxson

John Paxson started driving rocket cars in 1972. He drove the *Courage of Australia* rocket car until it was destroyed in a crash during testing of a new rocket motor. He also drove the *Armor All Rocket* for a couple of years, making over 300 runs between the two. John drove rocket cars on venues from coast to coast, setting over 25 track records before retiring. He obtained a top speed of 330.88 miles per hour at the Irwindale drag strip in California. The Armor All car was sold a couple of times, but John ended up buying it back and restoring it to its original condition and then donating it to the Wally Parks NHRA Motorsports Museum in Pomona, California.

Jack McClure

Jack McClure was not completely crazy, but he was definitely courageous and a true showman. He started out racing a Turbonique rocket-powered go kart in the mid-1960s. In the early 1970s, he had Glen Blakely build a new kart and Orville Porter built the 1,000-pound-thrust hydrogen peroxide rocket for it. Jack went on to make over 60 appearances, hitting top speeds of over 200 miles per hour.

Chuck Suba beside the *X-1* rocket car.

Chuck Suba

Chuck Suba was the primary driver of the *X-1* and was scheduled to drive the *Blue Flame*. Unfortunately, he lost his life in a Top Fuel dragster and was replaced by Gary Gabelich. Chuck's best time recorded in the *X-1* car was 265 miles per hour in 5.90 seconds. Chuck was not only a skilled driver but had also built a number of jet-powered dragsters.

John Allen Hudson

John Allen Hudson drove the *Texas Starship*, which ran 328 miles per hour in 4.80 seconds in October 1977. The car is now on display at Don Garlits' museum in Ocala, Florida.

John Allen Hudson is pushed to the starting line in his *Texas Starship* rocket car.

"Fearless" Fred Goeske

"Fearless" Fred Goeske ran a number of successful Funny Cars for many years before switching to rocket cars. I built a 5,000-pound-thrust rocket for his Monza Funny Car, which dominated the rocket Funny Car scene until it crashed from a rear-tire blowout. The car was repaired and put back on the circuit with a new driver. Fred was fortunate enough to continue on the circuit for a few years by buying several different rocket cars from other racers, who began dropping out when the hydrogen peroxide source dried up.

Fred Goeske mishap.

LEGENDS OF THEIR TIME

Jim Hodges with chutes deployed.

Jim "Gator Man" Hodges

Jim Hodges, the "Gator Man," began his 17-year drag racing career with a 1960 Chevrolet that he raced southern style. In other words, "Run what you brung." He soon learned there was no money to be made in that type of racing, so he stepped things up a notch and built his first flop-top injected Chevy Funny Car, which he affectionately called *Bonsai*. It proved true to its name when, in 1967, it won Jim the Canadian Nationals. He match-raced his cars all over the United States and won 43 out of 47 times.

In April 1973, Jim contacted me to build his first rocket-powered Funny Car. Sadly, when the car was about 80 percent finished, he was involved in a tragic firearm accident that left him with a badly shattered leg and was hospitalized for four months.

In Jim's younger years, prior to his phenomenal racing career, he'd made his living trapping alligators and selling their hides—hence the name Gator Man, given him by his friends. Jim was always quick to share his southern hospitality with a warm handshake and a friendly smile. Happily, he made it back out on tour in 1977 with his highly polished, bright, multicolored rocket-powered Funny Car. He and the car retired around 1979.

Tim Keselur's rocket-powered go kart.

Russell Mendez

Russell Mendez was the owner of the *Free Spirit* hydrogen peroxide rocket–powered dragster built by Glen Blakely and Arvil Porter with a 260-inch wheelbase, boasting a 5,000-pound-thrust rocket motor and weighing in at 900 pounds. Russell was one of the most world-renowned drivers there was, with a record run made in the early 1970s of 325 miles per hour in 4.7 seconds. An exceptionally nice person who'd do anything for anybody, he was tragically killed as his parents looked on while driving the *Free Spirit* at the sixth-annual NHRA Gatornationals in Gainesville, Florida, in March 1975. The accident occurred just one day after his 35th birthday.

Ramón Alvarez

Ramón Alvarez drove the *Free Country* rocket-powered go kart. The kart had a wheelbase of 52 inches, weighed approximately 200 pounds, and was powered by a 1,000-pound-thrust hydrogen peroxide motor. It was part of the Rocket Crusade Racing Team and had originally belonged to Jack McClure. The kart's top quarter-mile speed was recorded at 215 miles per hour.

LEGENDS OF THEIR TIME

Ky Michaelson, Ed Ballinger, NHRA official Bernie Partridge, Tony Fox, Jerry Hehn, and Vern Anderson get cars and drivers licensed at the Minnesota Dragways.

Ed Ballinger in the *Conklin Comet* races a U.S. Navy Blue Angel during a promotional stunt.

Chapter 18

Rocketman's Toys

Ky Michaelson's space-age rocket-powered wheelchair brings new meaning to the idea of retiring quietly.

I've spent a lot of time talking about past racers and racing experiences, but there's a whole other side to rocketry that I haven't touched on yet. I want to share some of my personal rocket-powered vehicles and experiences with you in hopes that you'll enjoy my creativity, maybe get some ideas, and go ahead and build some toys for yourself.

After retiring from the world of high-speed rocket racing, I continued building rocket-powered vehicles both as a hobby and for my own entertainment. My ultimate dream is to someday have my own museum so I can freely and publicly share my hobby with others, and I'm quickly accumulating enough toys to make that dream a reality.

Rocket-Powered Wheelchair

Now, back to the word "retirement." I was sitting in my office chair one day just wondering what it would be like when I got so old I'd need a wheelchair to get around in. Well, one thought led to another, and it wasn't long before I had this bright idea to build my own. I also knew that if I was going to continue being known as the Rocketman, it wasn't going to be just an ordinary chair; it *had* to be a rocket-powered one, complete with high-speed tires, disc brakes, and a parachute. All the extras were just in case I decided to make a high-speed run in front of my house—that and ensuring I didn't speed on past the local coffee-and-donut shop the cops hang at when there aren't any bad guys for them to chase.

Once I get an idea in my head, I just have to go for it. I wasted no time, and after scrounging around the local scrap-metal yards, I came up with enough surplus 6061-T6 aluminum to build what I thought was going to be the coolest mode of transportation a senior citizen could ever dream of. I decided to build what I would call a semimonocoque main frame. I used a seat off a new garden tractor and added a quick-release seatbelt so I wouldn't fall out if I happened to have a senior moment or hit a parked car while cruising down the road. Once I completed the frame, it was time for the power! I decided to build a 250-pound-thrust hydrogen peroxide motor to propel the thing, so I built a 2.5-gallon stainless-steel fuel tank, thinking I couldn't get into too much trouble with that amount of fuel, yet it would have enough burn time to definitely get my blood pressure way up.

I used two aluminum scuba tanks to hold the nitrogen to pressurize the fuel system. I dug through my old parts boxes and came up with a real nice pressure regulator that I scrounged out of a missile I found lying out in a scrap yard in California 30 years ago. (Oh, but those were the good old days—when the government was lax, and you could find surplus goodies that were still not taken off the top-secret list, but nobody knew their value, or even what the heck they were. You just brought them to the counter and paid by the pound. I once bought a guided missile out of an SR71 for $275; I later learned the government paid over a million dollars for it. Interestingly enough, I now realize that the only time I ever got a good deal from the government was at the scrap yard.) Anyway, I decided to empty my piggy bank and really splurge to buy a rack-and-pinion steering box for this high-speed wonder. Before I knew it, this rocket-powered wheelchair had turned into a sophisticated Indy race car. Well, I'd gone that far, so I decided I had to go all out and install a quick-release, aluminum, padded steering wheel.

After about four months of cutting, pounding, drilling, tapping, and assembling, my beloved rocket-powered wheelchair was up on its wheels. It was time to have my wife push me around the shop to see how well it steered. It handled great. After she went back inside, I just sat in the chair and wondered how it would feel to get real old and not be able to get around on my own. That's when I made yet another key decision: this thing needed headlights and taillights, in case my old age gets to a point where they won't renew my driver's license—at least I'd have my trusty wheelchair to sneak out at night for a little entertainment. When the chair was complete, I disassembled it and had all the aluminum and stainless steel polished. I don't have any rich uncles, so I had to reach deep in my pockets, to the tune of $2,000, to get my brightly polished parts out of hock.

After a week of carefully reassembling my new toy, I just stood back and looked at it with pride. It was done, and I thought it was a work of art, maybe even a masterpiece. But, there was one major problem. I called all the local chemical companies, and there was no 90 percent hydrogen peroxide available. All of a sudden, there was a sinking feeling in my stomach. How

was I ever going to run my rocket-powered wheelchair? A couple of months later, I received a call from a producer who worked on the television show *Ripley's Believe It Or Not*. Somebody had told them about my latest creation, and they wanted to visit Minnesota to produce a show on my very unique mode of transportation. Without hesitation, I told them to come on up, and they said they'd be there in a week. A *week*! Well, it was an opportunity I just couldn't pass up, so the wheels in my head started spinning, and I mean fast. I had to come up with a plan. I knew I'd never have access to hydrogen peroxide in such a short time frame, but I could buy a solid rocket and mount it on the back instead, so that's the route I took.

The film crew arrived, and I was ready. Now mind you, I never even test-drove this thing prior to their arrival, but I wasn't going to let that stop me. There's a long, long road with a gradual downhill grade and no crossroads about 2 miles from my house that had become the private testing spot for all my creations. That's where I took the film crew.

We decided to pull the wheelchair with a rope I'd be holding and get it going 45 miles per hour. Then I'd let go of the rope and fire the rocket motor. I very trustingly gave my wife the honor of pulling me with our pickup truck. I fastened the seatbelt and was ready to go. Just then, I got to thinking about how my wife had a very large life insurance policy on me. Before I had a chance to change my mind, however, the director shouted, "Action." We were off and running, and when my wife shouted out "forty-five miles per hour," I let go of the rope. I soon found out that the 37-inch wheelbase made the handling a little touchy. At one point, I oversteered and was soon up on my two left wheels. When I got it under control, I pushed the button to light the rocket. By the time I was halfway down the hill, I was going over 65 miles per hour in this thing.

I knew it was time to deploy my parachute if I stood the chance for a safe stop before the road curved sharply. The chute deployed beautifully, and it was a picture-perfect ride. The producer was happy with the shot, and I had just made the fastest run ever in a wheelchair.

Some time later, my good friend Bruce Lee volunteered to make some 90 percent hydrogen peroxide. His plan was to make a couple of gallons for my 65th birthday present so we could fire up the wheelchair at my party. I was a little concerned about his new adventure because in the past, everyone I knew who tried to make 90 percent hydrogen peroxide ended up blowing up their buildings or getting burned badly. Bruce was determined, though, and did a bunch of research beforehand. He decided the safest way to do it was to buy 30 percent hydrogen peroxide, put it in a Crock Pot set on low, and sit back and wait until the water evaporated. He hoped the end result would be a high concentration of the good stuff. In no time, his kitchen looked like my old chemistry lab in school, complete with Pyrex pitchers, beakers, a balance scale, hydrometers, a gas mask, and, just in case things didn't go quite right, a fire extinguisher. The good news is that Bruce didn't blow himself *or* his kitchen up. The bad news is when we tried using his home brew in the rocket wheelchair, all we got was a lot of steam and the majority of the peroxide didn't decompose properly. We came to the conclusion that either the catalyst pack was not effective or the peroxide was not at a high-enough percentage to properly react to it. Never fear, though, because the story doesn't end here. We're *still* cooking, and it won't be long before even Betty Crocker will be proud of our efforts.

Rocket Kart

By now you know that throughout the years I have built a number of types of propulsion systems using various fuels and propellants. As hydrogen peroxide had become nearly impossible to obtain, I decided to switch to an alternative fuel, something that was easily obtainable and didn't require a federal explosive license to have in my possession. I found the ideal combination that solved all those problems. The system I'm about to explain is a very basic one that's been used in model rockets and has also been used by some aerospace companies. It is called a hybrid rocket motor, and it uses nitrous oxide for an oxidizer. If you're wondering what is used for fuel, I hope you're sitting down. During my experimentation days, I used cardboard, plastic, fireplace logs, and, believe it or not, a 2-inch by 24-inch summer sausage! It is the ideal propulsion system and is safe to use if you are cautious and fully understand what you're doing.

The first hybrid vehicle I built was a go kart because I wanted to see if I could develop a throttle system that would allow me to turn the engine completely off and restart it. I also wanted to have a full throttle range. One of my favorite pastimes is attending swap meets, and that's where I found a well-used go kart for my experiment. Although it was in pretty bad shape, all I really wanted was the bare frame, so I disassembled it and scrapped the rest. Next, I sandblasted the frame to get rid of the rust and old paint. It looked almost brand new after that process, and I was very pleased. I machined a motor case to hold the plastic fuel grain and then mounted two scuba tanks on the kart, one to hold nitrous and the other to hold oxygen. I then mounted a high-flow solenoid valve on each tank so I could control the flow of the oxygen and the nitrous. The nitrous line has a stainless-steel ball valve between the solenoid and the motor, so I'm able to throttle the rocket motor if I want to. This system is the easiest and simplest propulsion system I have ever built.

ROCKETMAN'S TOYS

Jodi Michaelson can't let her husband, Ky, have all the fun.

I went to the local welding supplier and purchased a tank of automotive-grade nitrous and a tank for welding oxygen so I could test-fire the rocket before I polished and painted everything, in case I had to make any major changes. The rocket motor has a fuel injector to atomize the nitrous and a slug of PVC plastic for a fuel grain with a hole through the middle. At the end of the fuel grain is a converging/diverging nozzle made out of high-grade carbon. To run the motor, I fill both the oxygen and the nitrous tanks and then install an electric igniter in the motor, making sure the igniter goes all the way back to the fuel injector. At that point, the rocket is ready to run. After that? I just jump in, tell my loved ones how much I love them, and push the igniter button and the oxygen flow button. As soon as the motor lights, I open the nitrous valve, shut off the oxygen solenoid, and sit back and enjoy the ride. I highly recommend you drink at least three cups of black Colombian coffee first, though, so your adrenaline will already be a pumpin'. It just adds to the thrill of hearing and seeing a screaming, fire-breathing monster 2 feet from your head!

ROCKETMAN

Rocket Mountain Bike

I was shopping at Target one day when I spotted this really nifty bicycle. My wife, Jodi, is always telling me I need to get more exercise, so that was just the pretense I needed to buy the bike. Now, if you know me, you'd know that there's no way on this green Earth I'd ever actually pedal a bicycle or go on a bike ride. But I bought it anyway. It sat out in my shop for about a week, until I thoroughly convinced myself I really didn't need the exercise. Instead, I thought to myself, "Why pedal it when you can just put a rocket on it? Just think how awesome that would look." Ah yes, another toy for my collection: a rocket-powered mountain bike. Just like that, my mind was made up, and I set to work. I decided to build yet another hybrid rocket to propel me down the street at breakneck speeds and used a small scuba tank for my nitrous supply with another small tank off a splat-ball gun for oxygen. (Attention, Target shoppers!) I built this motor a bit differently from the last one, because this time I machined two blocks of aluminum to hold the PVC plastic fuel grain in place. Four threaded rods retain this system, so there is no motor casing. On each end of the fuel grain, there's an O-ring. When you tighten up the four threaded rods, the O-rings seal up against the aluminum block on one end and up against the carbon nozzle on the other, which prevents the hot gases from escaping. I used two NOS solenoid valves with two different port sizes, so I've now got a two-speed rocket because I'm able to control the nitrous flow. (I proudly call this is a two-stage system: fast and *faster*.)

Jodi lights up the night sky while zipping around on her rocket-powered mountain bike.

Jodi was into it from the moment it was done. As soon as I got it finished, she begged me to ride it at an upcoming event. Every year, about 300 gearheads show up at a local motorcycle shop in Minneapolis and have a tire burnout and wheel-standing contest. Needless to say, most of the people who attend this annual event are hard-core bikers who like their food and their beer. Well, I agreed to take the bike there, but I persuaded Jodi that *she* should ride it instead of me. (After all, I was still holding onto my antibicycle beliefs.) She agreed, so we went. By the time it was her turn to ride, the crowd had definitely gotten a little on the rowdy side. The announcer asked everyone to move out of the street and back onto the sidewalk. With everyone back, Jodi started pedaling toward the crowd. One of the 300-pound bikers, who looked like he was allergic to water (and therefore bathing), shouted out, "Chicken." Just then, Jodi fired the rocket motor, and it was absolutely awesome. The motor started with a loud roar and bright orange and red flames and lots of black smoke shot 10 feet out the back. It wasn't long before everyone had their hands over their ears to protect them from the high-pitched scream of the rocket motor. The bike became a crowd favorite and Jodi right along with it. She was stunned as people started asking her for her autograph! Yah, sure, you betcha', oh don't ya' know. We know how to have fun in Minnesota!

Twin-Engine BSA Bike

In 1960, I drag-raced motorcycles as a hobby and did very well in my class. But, as was usual back then, that wasn't good enough. I wanted to take an even bigger step and go after the big boys on their Harleys in the open class where anything goes, so I decided to build a twin-engine drag bike using two 650-cc BSA motors. I proudly named it *The Centipede*, because of its long length. My friends called it something else—*The Widowmaker*. That's because when I shifted the two-speed transmission, the rear

To this day, I've never fired it because I want it to stay as pristine as it is right now.

wheel would slide all over the track. One night, my friend Clem Larson asked me if he could drive it at Twin City Speedway. He had plenty of experience, so I let him take it down the track. Lo and behold, when he shifted into second gear, he somehow stuck his foot inside the rear wheel. He was pulled off the bike and dragged down the track with his foot stuck inside. It was a miracle that he suffered no broken bones, but he did have a lot of asphalt burn. The bike, on the other hand, didn't come out of the ordeal as well—it was totally destroyed.

Some 30 years later, I decided to build a new twin-engine bike for the museum I hope to have one day, and let me tell you, it turned out to be absolutely beautiful—so much so that to this day, I've never fired it because I want it to stay as pristine as it is right now. I've entered it at a couple of car shows, and it's done very, very well. After one of those shows, there were a number of people standing around me and the bike, and someone said to me, "So, Ky. What are you going to build next?"

I kind of looked around and paused for a little bit, trying to think of something I didn't have and fugaciously replied, "A rocket-powered toilet." Everyone just laughed. The thought was in my head, but I just sort of put it in the back of my mind. Well, about three weeks later, a guy knocked on my door and said, "I have something for you." I didn't know this guy from Adam, but he seemed really nice and was definitely anxious to show me something, so we went out to his car, and he opened the trunk. There sat a used, stainless-steel toilet. He said, "Here, it's yours."

SS Flusher, America's Secret Weapon.

What a surprise that was to me. Out of curiosity, I asked, "Where did you get it from?"

He promptly said, "I'm a plumber by trade, and I work in the prison system." Before I could say one more word, he immediately followed that remark with, "Don't ask me any more questions." I obliged, shook his hand, and thanked him before he drove away.

Rocket-Powered Toilet

Well? It was time to get to work. I mean, it's not just every day a stranger shows up at your door and presents you with a toilet, let alone a stainless-steel one! I decided if I was going to build this monstrosity, I was going to take my time and make it show-worthy. I picked it up and put it on my workbench. That's when it hit me. I immediately wondered if I should wash my hands every time I touched it. I mean, I knew it was used, and I also knew a lot of unsavory butts had sat on it in the past!

I decided to build this new project on four wheels and once again wanted it to be powered by a hybrid rocket motor. I had grander plans for this thing, though, because it was really going to be a one-of-a-kind creation, worthy of driving in parades or even around the pits at some of the local racetracks. So, with that in mind, I also wanted it to be powered by a small reciprocating engine. That way, when I wanted to feel the exhilarating power of a rocket, all I'd have to do was push a couple of buttons.

I built the frame rails out of aluminum and used high-speed go kart wheels and tires. I added disc brakes for extra stopping power and, just as I had with my go kart, used a rack-and-pinion steering box with a padded racing steering wheel with a quick disconnect. From there, I used two oxygen-breathing tanks for my nitrous supply and a smaller one for my oxygen source. This time, I used three NOS solenoid valves with three different port sizes for three levels of power—slow, fast, and hold onto your hat! I didn't want a wimpy rocket, so I built a 6-inch hybrid that had enough power to get my attention and everyone else's standing nearby. I added a couple of homemade missiles on each side to give it a military look and even had some fun by posting it to my website for a "name the toilet" contest.

It was promptly dubbed the *SS Flusher, America's Secret Weapon*.

This thing is not only unique in appearance but also has a very unusual starting mechanism. I push the "flush" button to start the oxygen flow, hit the igniter button, press one of the nitrous solenoid buttons, and then? I hang on.

ROCKETMAN

I had the thing polished to a mirrorlike luster and entered it in the Gopher State Timing Association (GSTA) Rod and Custom Show, where it took first place! In my mind, the *SS Flusher* is a fine piece of art and something that I'll treasure for a long, long time.

Rocket Luge

Over the years, I've heard many, many interesting stories, but one that stands out in my mind was that someone accidentally knocked over an oxygen bottle, causing the valve to come off. The pressurized oxygen shot out so fast it threw the tank straight through a concrete block wall. Stories like that, whether they're true or not, always get my thinking juices flowing. When I heard that one, I envisioned what it would be like strapping one of those high-pressure tanks to something and going for a ride. In reality, that would be pretty death-defying, but then again, if you could somehow control that power and be able to steer—what an absolute thrill ride *that* would be.

ROCKETMAN'S TOYS

This is something Buck Rogers would have liked to use back in the 1950s.

It wasn't long before I was digging around in my shop. I wasn't exactly sure yet what it was I was looking for, but I had a vision of a really awesome vehicle etched in my mind, and I just kept looking. I found a high-pressure solenoid valve with a 1-inch port that would allow me to control thrust. My next step was to buy two 6-feet by 6-inch high-pressure tanks. I decided I'd fill one of the tanks with about 15 gallons of water and then pressurize it with 3,000 psi of nitrogen. I knew by adding water, it'd give me some extra mass, resulting in even more push. I'd just fill the other tank with plain nitrogen. Once I had the propulsion figured out, my vision came to fruition. I proceeded to build an all-aluminum, low-slung chassis with rack-and-pinion steering, disc brakes, and a Rocketman chute for extra stopping power. I called it a rocket luge.

This toy has about 2 inches of road clearance, and instead of sitting on it, you are almost completely lying down. It looks and feels like a strap-on rocket. I did a lot of polishing on it and had the whole luge anodized a beautiful purple. A film crew from Germany came over to do a show on me just as I was

putting the finishing touches on it. I told them they'd come at just the right time because it turned out that I had made arrangements for the local police to block off a road so I could test it the following day. That night, I got to thinking about all the things that could possibly go wrong, and the worst fear I had was that all the stories I'd heard were true. What if the luge could in fact take off and fly? Would I be able to hang on, and if not, would I be thrown right off the back of it? I didn't like either scenario; they both felt like they could be very harmful and, worse yet, very painful!

The next morning, I got up early and decided I'd better test fire this thing before the film crew arrived. I put it into the back of my pickup truck and strapped it down. Then I put 15 gallons of water in it, loaded the tanks with 3,000 pounds of nitrogen, and test fired it.

When I pushed the solenoid valve button, it came alive: it was like an overpressurized fire hose. The water just screamed out the back end. I am sure if my truck hadn't been in park, it would have pushed it. That's all I needed to see; everything seemed to work perfectly.

As soon as the film crew got there, we headed down the road to my test spot. The police had the road blocked off, so I didn't have to worry about any cars in back of me. I strapped on my helmet and my leather jacket, just in case things didn't go as planned. You see, at my age, I've become allergic to asphalt rash and pain! When I lay down on the luge in race position, it felt like I became a part of it. I found myself thinking, "This is something Buck Rogers would have liked to use back in the Fifties." Once again, I entrusted my wife to pull me with a rope attached to the back of my pickup truck (there goes that life insurance policy thing again) so I could get the feel of the luge before I started the rocket. For some reason, she's always more than willing to do these things for me, and she got me going about 30 miles per hour. I let go of the rope, drove off to the side of the pickup truck, and smiled at Jodi as I pushed the solenoid button. All of a sudden, I was slammed up against the headrest. The luge took off like a bat out of hell.

I took my finger off the solenoid valve for a second and then hit it again. This time, I pushed it for about 4 seconds. I didn't know how fast I was going, but I knew for certain it was a heckuva lot faster than I wanted to be going, so I deployed the 7-foot parachute and slowed down enough to come to a safe stop before coming to the end of the road. The amount of power in this glorified squirt gun had really surprised me.

Somewhere down the road, I'm tempted to build something else using this same propulsion system because it is fairly safe and relatively inexpensive to build and run. If you are as much of an adrenaline junkie as I am, this is definitely the toy for you. Just remember one thing: any time you're working with high pressure, *be careful!*

Rocket-Powered Flexible Flyer Sled

One winter day, I was in a hardware store and I happened to spot a small child-sized Flexible Flyer sled that was about a foot long. I recalled having a full-sized sled just like it when I was a small boy. My impulses took over once again, and I ended up buying the thing. As soon as I got home, I headed straight to the shop with it. My wheels were in fast motion: I was going to convert this kid-sized toy to a man-sized one! I scaled it to 8 feet and then headed back out to pay a visit to the local tube bender and had him bend 1 1/2-inch aluminum L-channel runners. I made the articulating steering just like a conventional sled and then used hard oak that I stained and covered with two coats of polyurethane as the wood. The handlebars and speedometer came off a Harley-Davidson motorcycle, and the housing came from a historical Aerojet JATO solid-fuel rocket I had just never found a home for until then. That would be more than enough to power my giant Snow Flyer.

For many years, I had heard many variations of the urban myth about the guy who put a JATO bottle on his 1962 Chevy Impala. No matter which version of the story I heard, the ending was always the same. There were skid marks, brakes smoking, and someone going off a cliff or into the side of a mountain, resulting in the driver's death. The reality is a JATO only has about 1,000 pounds of thrust for 14 seconds—not nearly enough to get a 3,500-pound car going so fast that its standard brakes couldn't easily stop it at any time during the thrust duration. To prove my point, here's some land speed history for those of you who don't know it. In 1964, Walt Arfons and his crew built a car called the *Goodyear Wing Foot Express* that was powered by 25 Aerojet JATO solid-fuel rockets. There were 15 mounted in the rear and 5 on each side. On October 22, 1964, Bobby Tatroe drove the car at the Bonneville Salt Flats and, by firing the rockets in sequence, was clocked at 476.6 miles per hour one way through the flying kilometer. Then in the flying mile, the car ran about 580 miles per hour before running out of fuel while being timed. At that time, one JATO cost about $1,000, so each time they ran the car, it cost them $25,000. The car never did break Craig Breedlove's land speed record, but it was a gallant try.

ROCKETMAN

My JATO was left over from that venture. I made an adapter so I could either use a number of manufactured high-powered rockets or build one myself. I simply mounted a battery to the sled for an ignition source. We decided to run it out on the snow-covered sidewalk in front of the house. There's also a skating rink just down the hill, so we figured Jodi, who volunteered to go first, could make a slight turn and head out onto the ice. For the trial run, I mounted a K550 motor on it.

It just so happened that there were a number of skaters that day, and they had no idea what we were planning to do, so it was really funny watching their reaction to this gigantic sled. As soon as Jodi hit the ignition, they all turned around and covered their ears as she whizzed by in a cloud of smoke and fire. It looked like so darn much fun I just had to give it a try. But, before I did, I had to make one slight modification. I took it into the shop and quickly mounted a longer-burn-time K700 motor on it, and it actually pushed me to about 30 miles per hour. It was a really awesome ride and a lot of fun for everyone who was there. Most important, no one got hurt, and I just can't think of a better way to take the monotony out of winter. Now I've got a sled with a rocket to get me *up* a hill, and then I can just turn it around and head right back down full-speed ahead. Now, that's what I call innovation at its best.

Jodi rockets down the snow-covered sidewalk in front of our Bloomington, Minnesota, home.

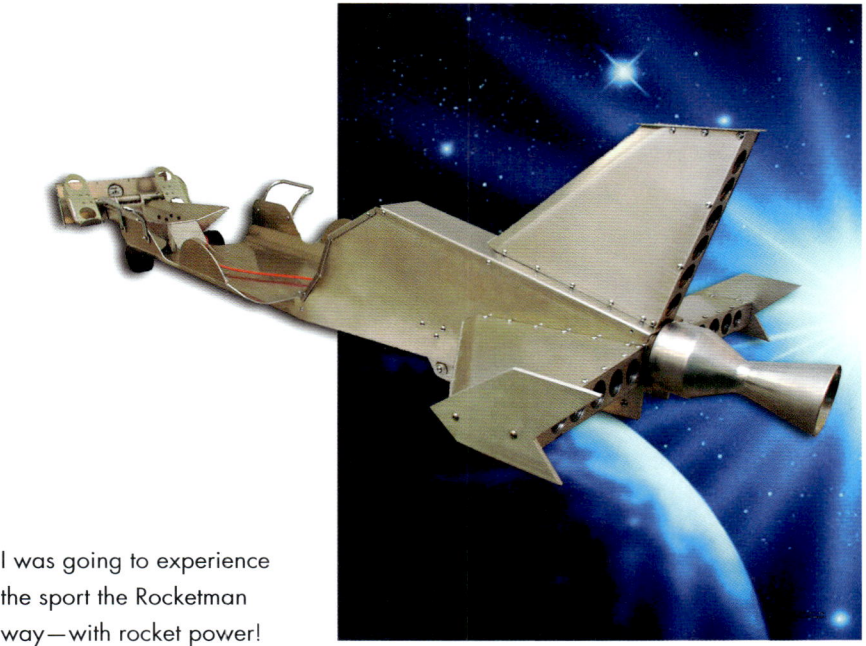

I was going to experience the sport the Rocketman way—with rocket power!

Rocket-Powered Street Luge

In addition to my rocket luge, I advanced to bigger and better. Downhill street luge racing is very popular out on the West Coast and in a few other states that have mountains and long, steep hills. Unfortunately, Minnesota has neither, with the exception of some ski resorts packed with skiers and not suitable for luge racing. Nonetheless, I found myself really intrigued by the sport itself and decided to look into it further, just for the heck of it.

I hopped on the Internet and did quite a bit of research on the subject, amazed to discover how crude the current technology in luge design really is. In fact, I concluded it was about as far from high tech as I'd ever seen, with prices for racing luges ranging anywhere from $500 to $1,800. The more I studied the pictures, the faster the wheels in my head spun. Research aside, I decided that hills or no hills, mountains or no mountains, I was going for it. I was going to experience the sport the Rocketman way—with rocket power! I'd learned about all the crude items out there, but I had a different idea altogether. I knew there was a lot of advancement in wheels and bearings, so I decided to use skateboard wheels and wound up buying the best wheels, bearings, and trucks available. The available chassis weren't much to look at but at least gave me something to go on.

I built an all-aluminum semimonocoque frame and then set to designing a braking system. (Oddly enough, most luges don't have brakes; the riders just literally drag their feet to stop, and although I'm sure the shoe companies are *very* happy with that arrangement, it is much too primitive for me.) My system is about a step above the Barney Rubble class, but hey, it actually works pretty well. It's definitely better than nothing. I used a piece of rubber from a truck tire and attached it to a camlike mechanism that is controlled by a foot pedal. It gives you a slow but safe stop, minus the hot foot!

As for the rocket power, I decided to use solid-composite propellant motors because the propellant is very similar to what's used in a space shuttle. The main motor is a long-burn K700 motor. I've never determined the top speed of this street luge, but I've no doubt it could easily go over 150 miles per hour, which, given the fact that you're lying flat on your back only inches from the ground, is *fast*! The ultimate speed would just depend on how steep the hill is and how fast you were going before you fired the rockets; sort of like flying without an airplane.

Rocket-Powered Outboard Motor

I live on a small private lake, and one of the things I really enjoy in the summer is to jump in my canoe in the evenings and paddle around to take in nature. It's relaxing, but sometimes it gets relaxing to the point of boredom for me. Must be because there's just no fire or smoke or sound to it. Well, that being said, you guessed it. To keep myself from being bored and also give my neighbors even more to talk about while they're out enjoying their hot summer nights, I constructed an all-stainless-steel rocket-powered outboard motor for the canoe. I used the hydrogen peroxide propulsion system off of my son's rocket backpack and am able to literally clamp the motor to the back of the canoe, just like a conventional small outboard motor.

Being able to bail out any time by just jumping overboard gave me security in believing the thing was relatively safe. On our first voyage, we knew right off that we'd never break any speed records, but hey, it sure was entertaining watching the white, bubbling steam coming out of the water as the rocket thrust pushed us forward. While not the most practical thing I've ever built, it sure is an exercise in having fun for all. If nothing else, too, I'm sure the fish enjoy the extra oxygen we put into the water.

ROCKETMAN'S TOYS

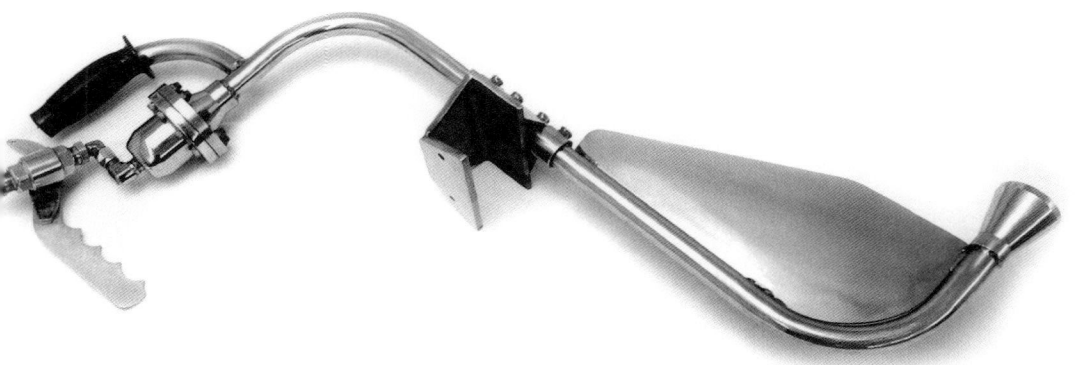

Buddy "Rocketman" Michaelson's rocket-powered tricycle.

Rocket-Powered Trike

Every year there's a gigantic swap meet at the Minnesota State Fairgrounds, and it seems like there's always something for everyone. It's an event I never miss and one that lends to my overwhelming desire to just buy everything in sight. That being the case, I never made it out of there without loading my pickup with goodies that I just couldn't live without.

ROCKETMAN

Buddy with his toys. He thinks it needs more power.

On one rendezvous, I found a child's three-wheel tricycle that probably dated back to the 1930s. Now, you might ask what an old guy like me would want an old trike for, and the answer is simple. As I stood there admiring it, I found myself thinking, "Whoever built this thing forgot to put a rocket on it." That was it. The trike was coming home with me, so into the pickup it went. I no sooner got home than I had it unloaded and up on my bench, where I stood back and admired the old relic. It had a real classic look, and I soon found I just didn't have the heart to start hacking the thing apart. I decided it really was a part of history that needed to be preserved, so that night I lay in bed thinking that instead of destroying that one, I'd use some of the same design concepts and just build a brand-new one from scratch. As I lay there, I also got to thinking about my four-year old son, Buddy. I hoped that someday he would follow my footsteps and become interested in designing and building rockets and maybe even become an astronaut.

When I awoke the next morning, I had a new vision. The time had come, so I told my son that together, just he and I, we were going to build a rocket-powered trike and enter it in a car show. His eyes got big, but I'd saved the best for last. I then told him that when the car show was over, the trike would be all *his*! You should have seen his eyes light up then. I knew this would be a great learning experience for both of us, because in all honesty, I'd never worked with a four-year-old child before, but I knew he was more than ready to start, and so was I.

ROCKETMAN'S TOYS

In order to get some ideas, we went down to the local Target store to look at bikes. I spotted one that had the components we'd need, so I bought it and brought it home. Little Buddy was just ecstatic that good old Dad had just bought him a new bike . . . that is, until I walked over to the band saw and cut it right in half. With eyes as big as saucers, he stood there in silence with a surprised look on his face, totally clueless as to why I had just done that. At that point, I sat down with him and explained that we were going to use the handlebars, front fork, wheel, chain, and sprocket on the new rocket trike. It didn't take me long to realize what a valuable tool this project would become as a learning experience for him.

I made it a point from then on to sit down with him and explain every process and how things worked before I began and to make sure he fully understood why I would build things the way I did. It brought pure joy to my heart to see him so willing to participate and to ask questions and to run for tools, eager to help in any way he could. I even let him help me when I machined parts on the lathe.

It didn't take me long to realize what a valuable tool this project would become as a learning experience for my son.

The trike is powered by a 4-inch by 24-inch hybrid rocket motor and has considerably more power than the rocket bike. We used two front wheels off a junior dragster for the rear wheels and a mountain bike wheel on the front with stock scrub brakes. I used disc brakes off a go kart on the back, just in case, once again, it ever has to come to a screeching stop. It took us about three months to build, and we finished it just in time for the big Super Bowl party. I had hoped to run it out in front of the house, but the weather decided not to cooperate. We wound up having a big snowstorm, and the temperature dipped to 10 degrees below zero, so we had no choice but to wait. I told Buddy that even though we couldn't run it, we could still test fire it, 'cuz that's what all guys who build rockets do. He settled for that, so we gathered some of the party guests and headed out to the shop. I got everything set up, and you should have seen the pride in my son's eyes as the flames shot out the back of his rocket-powered trike with an ear-piercing roar.

Spring came, and in April, we entered it in the 48th GSTA Rod and Custom Show and won a first-prize trophy and $50 in the special-interest class. That project was truly a heart-warming experience for me and has brought my son and me closer together. I was finally able to let him witness firsthand what it means to take an idea and turn it into a workable piece of hand-crafted art. I'm sure that trike will grow old right alongside Buddy.

This is another rocket-powered snowmobile I built for Lee "Sno Ball" Smith. After Lee passed away, his wife sold it back to me. I repainted it and polished all the stainless steel and aluminum.

Rocket-Powered Snowmobile

Minnesota is known as the land of 10,000 lakes, but in the winter we become the land of 10,000 frozen lakes, boasting over 250,000 registered snowmobiles. I know that sounds impressive enough, but guess what? Of that number, there were no registered rocket-powered snowmobiles. That is, until I decided to change that statistic with a little ingenuity and work. After all, I have a flair for those kinds of things, having built three total throughout the years. Two of them I built for a friend of mine whom we affectionately called "Sno Ball" Smith. The smallest one was an Arctic Cat Kitty Cat built solely for the fun of it.

I built a 250-pound-thrust hydrogen peroxide propulsion system motor for it, figuring Sno Ball couldn't get into *too* much trouble with it, because he weighed well over 200 pounds, and the sled could only go about 15 miles per hour. However, that didn't turn out to be the case.

One day at a snowmobile race, he was proudly showing off his snowmobile to the other racers. Now one of the secrets on how to keep warm on a cold winter day up here in Minnesota is to always carry a flask of rum with you and then be sure to say things like, "Yeah, you betcha," or "Yeah, don't ya' know." Well, Sno Ball was a true Minnesotan in every way and always carried a flask of brandy with him. This particular race day, he'd actually had a couple of flasks before going out in the cold, so let's just say he was a bit on the tipsy side for the occasion. He fueled up the snowmobile with 2 1/2 gallons of 90 percent hydrogen peroxide and, while standing alongside the sled,

Sno Ball Smith on his rocket-powered Arctic Cat Kitty Cat snowmobile.

began explaining to the crowd that had gathered how the system worked. He'd captured their attention, and they were all really intrigued, so he went on to explain how the throttle worked. It would have been a real good idea for him to do so while sitting on the sled, but he was still standing beside it when he reached down and pulled the throttle ball valve handle. The sled moved slightly forward and, without realizing it, his hand slipped and accidentally hit the throttle wide open! The thing took off like a bat out of hell for a short distance, until it hit a snowbank and launched up at a 45-degree angle, setting the world snowmobile jump record with no one on it. It was quite a sight to see and just reaffirmed my feelings that this guy was truly an accident looking for a place to happen.

As I was writing this book, I was going through some of my old newspaper clippings and found an article with the headline "Flaming Truck Like Bad Dream." You guessed it—that flaming truck belonged to none other than Sno Ball Smith. He made his living sharpening saw blades, and he and his wife had been out in his pickup truck with a camper on top making a delivery to one of his customers. They were traveling down University Avenue, which is a main drag in St. Paul, Minnesota. Without thinking anything of it, he had a couple of containers of 90 percent hydrogen peroxide in the back of the truck, alongside the saw blades. While that wouldn't normally have been a big deal, unbeknownst to him and his wife, the sharp blades had been rubbing up against the plastic containers for a good long while and had actually worn a hole in one of them. *Not* good! The peroxide poured out and started a fire in the back of the truck.

Sno Ball smelled smoke and looked in his rearview mirror and saw flames shooting from the back, later describing it as a "tornado from hell." He screeched to a halt in the middle of the road, and he and his wife bailed. He forgot to put the truck in park, though, so it rolled straight down the block and directly into the front display window of an Oldsmobile dealership. Luckily, the back of the truck remained outside the building, because the collision was followed by a *huge* explosion! An eyewitness was quoted as saying, "There was debris flying all through the air. There were parts of the truck that went higher than the Montgomery Wards building nearby, and that's probably 150 feet." A 21-year-old car salesman said there were about 10 customers in the showroom at the time, and as soon as he saw the truck coming across the intersection, he just ran into the office and hit the floor. A St. Paul fire investigator said it appeared that the hydrogen peroxide leaked out of the container and may have mixed with dirt and oil in the truck, oxidizing and building up heat to the point that it triggered itself. The official called hydrogen peroxide a "highly unstable liquid" but didn't believe a permit was required to haul it in the city.

I just thought I'd throw that story in so you'd understand that working with hydrogen peroxide can be extremely dangerous if you don't handle and store it properly.

I truly hope that by providing some of the insight from my past 50 years of experimenting and building rocket-powered vehicles, I've managed to inspire a few of you to perhaps follow in my footsteps and keep the space-age world alive and kicking when I'm no longer able to. It's really a magical feeling to be able to build such incredibly powerful things using your own two hands. I highly recommend giving it a try—carefully!

Gizmo

Over the years, I have had visitors come into my shop and say, "Ky, this stuff you build is like art." To me that's a big compliment, but to be honest I have never looked at my creations as art. I like to think it's more of an extension of me than art. Or, as I have always said, they're just my toys. Well, I eventually decided to build something that in my own eyes was art.

Jodi stands next to *Gizmo*.

ROCKETMAN

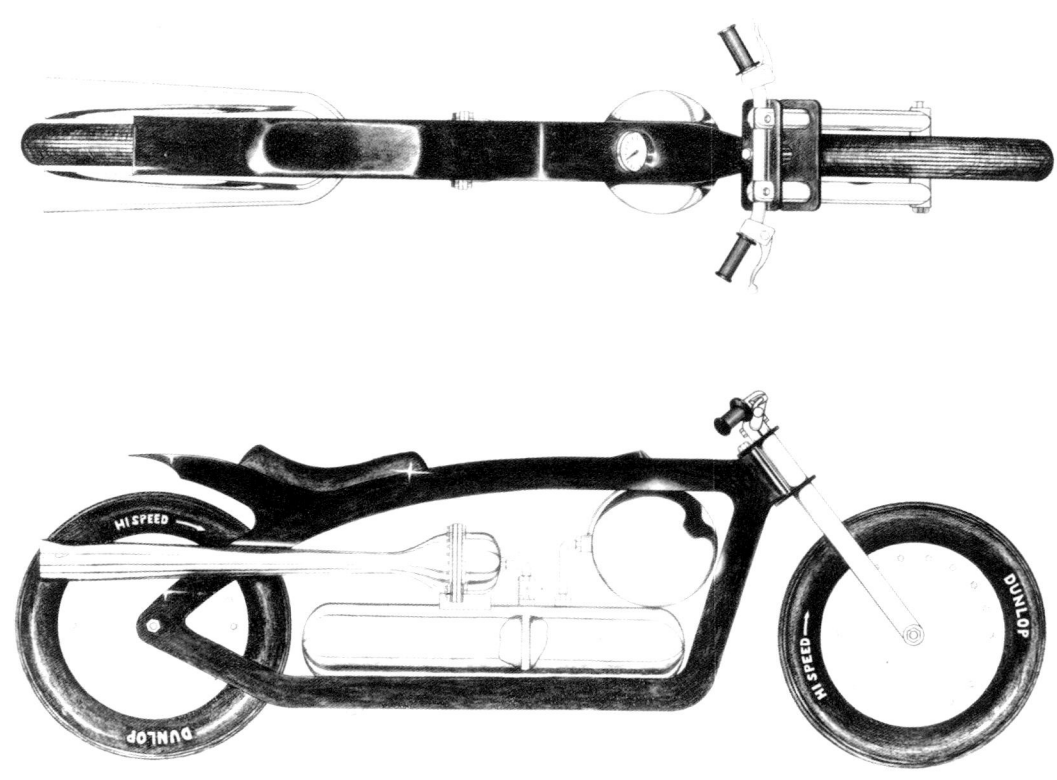

About 25 years ago, I designed a rocket-powered motorcycle that was made out of flat, lightweight pieces of aircraft T6-6661 aluminum. That design remained in the back of my mind for all these years, so I decided to build this unique design and kick-start my career as an artist.

In the past, I built a rocket-powered bicycle using a standard 24-inch bicycle. As my first attempt at building something artistic, I thought I should build a bicycle from scratch. My friend Bill Wilson donated a pair of beautiful aluminum wheels to the project. Many years ago, I saw a show on television called *Gizmo*. The show was about unique and weird creations, so I named the bike *Gizmo*. I spent hundreds of hours sawing, drilling, and fabricating the bike. The front forks are about as weird as you can get, but it's artsy, and I guess that's what art is all about. I built two 75-pound-thrust rocket motors to get some extra go out of *Gizmo*. When the project was done, I took it apart and had Joe Deters put his magic polishing touch on it.

Over the years, I've found out nobody on Earth can polish aluminum like Joe can. I am extremely happy with *Gizmo*.

I am also blessed to have the God-given talent to be able to build a project like this with no drawings or blueprints. Now I have a challenge: To try to out-do myself when I build my next project.

Rocket Chair

The newest creation to come out of my shop is this rocket chair. In 1965, Bell built a rocket chair using a standard chair out of the Bell commissary (pictured here). I have come up with an innovative rocket chair that has four nozzles for added stability. The chair has a 500-pound-thrust rocket motor on it and carries 7 gallons of hydrogen peroxide. The rocket chair will have a flight time of over 30 seconds and will be easier to fly than the standard rocket belt

The Bell rocket chair.

ROCKETMAN

ROCKETMAN'S TOYS

Ky models the rocket belt.

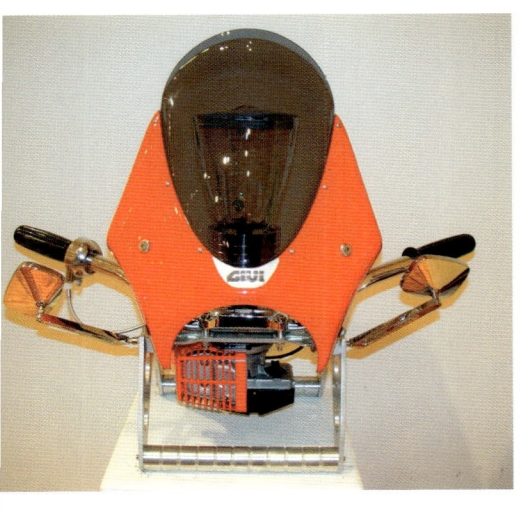

Ky's rocket-powered blender.

Part II
Space Shot on a Shoestring:
The CSXT Go Fast Rocket Becomes the First Civilian Rocket to Reach Space

The first part of this book documents the many amazing personalities I have had the good fortune to know and the creations I have built, worked on, and learned from, pursuing rocket-powered thrills and records on land and water. In addition to the fun and challenge they promised, these experiences arose from a desire to do what had not been done before.

While each of those projects presented difficulties and uncertainties, none was as taxing, consuming, or ultimately as thrilling as the quest described here: a seven-year adventure chasing a dream every rocket fan that ever was can relate to—to build and launch the first amateur rocket to leave Earth's atmosphere and enter the world beyond ours . . . space.

Chapter 19

Space Shot 2000

If great things come from humble beginnings, we were off to a good start. My Civilian Space eXploration Team (CSXT) needed government approval to send up a rocket of this size—which would cross airplane flight paths and be tracked by radar systems—and we were getting nowhere.

Then I received a call from one Jerry Larson, and his call literally changed the entire direction of my space program forever. Jerry worked for Lockheed Martin and was the lead flight designer for the Athena program. I couldn't quite figure out why this guy was calling me, but he quickly told me he'd actually heard about all the trouble I was having trying to get my program licensed. Then he dropped the most exciting news on me since I started my quest for space: He told me he worked with the office of Commercial Space Transportation (AST) on a regular basis and that he actually helped file the flight plan for the Athena rocket. I didn't even know what to say. He seemed to detect my apprehension and immediately assured me that he could answer any technical questions the AST office threw at me. He continued by telling me he had access to a program called the Sixth Degree of Freedom. Here I had been as close to a near standstill with my space launch as possible, and just like that . . . along came Jerry.

After several conversations with him, I came to realize where I had made my biggest mistake. It really had to do with the project's name I had chosen: the Civilian Space eXploration Team. I had used the word "team" inappropriately, because in all honesty, there was no team at all, other than my wife, Jodi, and me. We were it, and I had been foolish enough to think she and I could pull this off and make it on our own. Well, Jerry very quickly informed me that an effort such as this would require team players and lots of teamwork too. He immediately came on board as a very active and important team member himself and actually became our program manager. All I can say is, "Wow." In no time at all, he had completely turned things around for us and jump-started the whole program.

He got on the horn to our friend, Randy, the government contact we had not gotten too far with, and stated, "I'm now a member of this team and have as much desire as Ky does to not only make this attempt as civilians to launch this rocket into space, but to do so this year."

Randy retorted, "Well, in order to do *that*, you'll have to provide a million-dollar insurance policy to the Bureau of Land Management."

Jerry shot back, "Bring it on. You send me the papers, Randy, and we'll get you what you need." Just like that. I couldn't believe it. After all my months of endless conversations with the guy, Jerry persuaded him with just one phone call.

Things began moving with speed and precision, and we immediately renamed the project "Space Shot 2000." If Jerry seemed urgent to move things forward, he had good cause. Two organizations—the Space Frontier Foundation and the Foundation for the International Non-governmental Development of Space—had created a challenge. They would pay a $250,000 "Cheap Access to Space" prize to the first team to put a privately developed rocket into space by November 8, 2000. Our new goal became to get our rocket launched before the CATS prize deadline. In no time at all, we received the necessary paperwork from Randy and Jerry filled everything out, copied it all, and sent the originals via certified mail to both the Associate Administrator for Commercial Space Transportation (AST) and the Bureau of Land Management (BLM).

Our 1995 launch attempt.

Moments later, in 1995.

What remained of our 1995 rocket.

1997 rocket launch attempt.

The rocket coming out of the tower in 1997.

Paperwork aside, it was time to do the design work. Jerry lived in Colorado, so the two of us were on the phone together every day for a period of about six months to design our new rocket and its support equipment. I have a lathe in my shop at home, but the nose cone on this thing was so large that I ended up having to outsource a lot of the machine work because my lathe just couldn't handle it. The nose started out as a 250-pound block of solid 6060-T6 aluminum, and took 130 hours of precision machining to complete. Believe it or not, once fully machined, it weighed a mere 10 pounds.

While I did this machining work, Jerry handled technical details. One of the requirements we had to meet prior to the launch was to measure the wind speed at various altitudes to allow us to calculate the angle of the launch tower, so the rocket would land down range in a specified area. Jerry and his son, Tyler, worked diligently on designing the equipment necessary to make the calculations. We had to prove to the AST that even if something went terribly wrong, our rocket would land in the prescribed area. Jerry was a real trouper and toughed out a grueling seven straight hours with a computer program that helped with the calculations. I couldn't help but marvel at him, because if it had been up to *me* to figure out, I would have been using my 10 fingers and toes, and it would have taken me at least seven or eight weeks to complete. He got it down to a science, and it took 1,000 simulations to convince the AST we could do it.

While things didn't always go smoothly, we definitely remained in forward motion. As Jerry worked through all that fancy computer stuff, I got busy doing what came naturally to me: obtaining publicity and television coverage. I wanted this historical event to be televised at a national level, but I didn't want to give the rights away or have some production company come out with a show depicting the event as some kind of daredevil act or stunt special. The only way to avoid either of those things was to enter into a partnership, which is exactly what I did. I found a production company willing to produce a documentary and call it "Space Shot 2000." I entered into an agreement with them, with the condition that I be allowed to have input into the show as well. They even agreed to do a segment on some of my childhood adventures, which I was really excited about. I had always worked on shows and things reflecting other people's lives, so the prospect of having my own life captured on film was really a big deal for me. All the pieces came together, and the excitement level skyrocketed . . . until the Friday before the scheduled launch.

We had sent all our paperwork in for the launch license months earlier, yet when I called Randy that Friday to ask him where our actual license was, he told me he had no idea. In an instant, it felt as though my whole world had dropped out from under me.

I said, "What do you mean, you don't know where our license is?"

He started with his usual runaround, nondescript, indirect answers, and I just blew a fuse. I could no longer contain my temper. I screamed into the phone, "Randy, how does it feel to work all your life and never personally accomplish one thing? You have *no* backbone, and all you seem to care about is what happens to *you*, and what the files around that office say about *you*."

He fired back at me, "Listen here, Buster . . ."

SPACE SHOT 2000

Before he could say one more word, I fired right back, "*No*, you listen to *me*, Buster! We have done everything your office has asked of us and provided you with everything you have required! I have entered into a *contract*, Randy, with a production company to film this launch. Right now, as we speak, the truck is jam-packed full of camera equipment, and our five-man camera crew is leaving Minnesota Sunday morning to head to Black Rock. The rest of the crew is flying out on Tuesday. I have *got* to have that license. We were all counting on you to come through for us."

There was silence for just a moment, and then he said, "Call me Monday," and hung up the phone.

I know my conversation probably sounds mean-spirited, but he had finally pushed me over the edge with his continued arrogance and evasiveness. I couldn't sleep all weekend. I had a very small margin of faith that at launch time, I'd actually have my license in hand, but most of my rocket friends said it would never happen. Sadly, and for the first time, I was almost beginning to agree with them. But, a plan was a plan, so the five-man crew wished me luck as they climbed into their vehicles on Sunday morning and headed for Nevada.

With the crew en route, I very anxiously picked up the phone first thing Monday morning and gave Randy a call. "So, have you been able to find out where my license is?"

That's when he dropped the next bombshell. With a slight air of pure satisfaction in pulling the rug out from under me, he snarled, "Sure. I know *exactly* where your license is . . . it's in our attorney's office in Washington, D.C. This matter is now clearly out of my hands, so you'll just have to wait to hear from them."

Before I could even reply, I heard a click. End of conversation. I hung up, and the sinking feeling of helplessness, combined with rage, engulfed me. I couldn't believe this had happened. I just knew for certain that any attorney reviewing my application would find at least 100 reasons to deny it, and that's where the anger kicked in and took control. I took Jodi by the hand and told her we needed to have a very serious talk. We sat at the dining room table, and I shared all my feelings with her.

"Jodi, more than anyone else in my *life*, you know what this launch means to me."

She nodded in acknowledgment, so I continued. "I have not been issued a legal license to do this, Jodi. They are stopping me at every turn, even though I have given them everything they've asked for. We have invested so much time, energy, and money into this mission that I can't stand to see it abandoned."

"Well, Ky? What do you want to do now?"

"Let's get our bags packed today, and let's go for it. The crew is already on the road, I've got contractual obligations, and I just can't abandon this mission. I can't even begin to guarantee you when we'll be able to return home, but, Jodi, license or no license, I *will* be pushing the launch button."

The repercussions and risks of my doing this were immense. I stood the chance of encountering substantial legal ramifications and, worst of all, being fined up to $100,000. Jodi sat silently for a few minutes and then got up slowly from the table, pushed in her chair, and said, "How many team shirts shall we pack?" My wife is one of the most understanding people in the world, and she came through for me, even though this meant hiring a nanny and leaving our children behind for an unknown period of time. We worked together feverishly to get everything lined up, and our nanny arrived in no time. Everything was in place, and there was no turning back.

We headed to the airport with the rest of the crew members as planned. The flight brought us into Reno late Tuesday night, so I contacted Bruno's Country Club to let him know we'd be late arrivals. He was, as usual, more than accommodating and told me he'd wait up for us to be sure we'd get keys for our rooms. We made the two-hour drive, and he was true to his word. Not only did he have our keys for us, but he also gave the entire crew a drink on the house before we retired for the night. We had a quick breakfast in the morning and headed out to the desert. Jerry, Tyler, and the rest of our crew were already there, and the spirits were very, very low. I had been hoping that Jerry would maybe have some good news, and he was hoping I would, but that wasn't the case at all. We all gathered around to discuss our options. In no time at all, a unanimous decision was made to proceed as planned, with or without word from Washington. We all knew the ramifications but were determined not to let the government's lack of enthusiasm and support prevent us from trying to achieve our goal. The film crew immediately began shooting footage, which pretty much consumed our first day on location, and though thoughts of hearing from Washington were foremost on everyone's minds, the call never came.

Thursday morning, we got very busy and assembled the launch tower and the rocket. Our payload was a bit on the unusual side, as we had received a very special request from one of the cooks at Bruno's. When he found out we were sending up a piece of Bruno's famous ravioli as a joke, the cook approached me with his question. He told me how it had always been his father's dream to fly into space one day, so he asked, "Would you be willing to let me put some of my dad's ashes on board for the flight? I know it's a strange thing to ask, but would you?"

I could clearly see how much this meant to him, so I said, "Sure. Why not?"

In addition to the ashes we carefully placed inside, I had also printed up about 50 small postcards with our Space Shot 2000 emblem and team member information, postmarked out of Gerlock, Nevada, along with 200 silver coins. These items would be sold as commemorative, historical collectors' items, having flown to space and back. With the payload in place, we assembled the fin canister, loaded the fuel grains, and installed the nozzle. The entire team struggled to lift the tower into place, and with great effort on everyone's parts, we finally managed. Another full day, and a lot more footage was filmed. It was time to head back to Bruno's to see if anyone had heard anything.

Nothing yet, but Bruno tipped us off that any hopes of communications should really be handled through the Limited Liability Company in town, and that I should hook up with a gentleman by the name of Danger Ranger. I inquired as to where that office was and asked why he felt this guy would be able to help us, and I soon learned something new. In 1992, Danger Ranger founded the Black Rock Rangers, an institution patterned on the Texas Rangers and their historic role as guardians of a dispersed frontier society. Bruno went on to tell me that within the Black Rock Desert area, Danger is known to his friends and coworkers as Michael Michael, and that he joined the project in 1990 and oversees the security and survival of the Burning Man community. I then asked Bruno what project he was talking about, and what the heck was a Burning Man community? I had never heard of such a thing. One of the biggest kicks I personally get out of life is that no matter how much older I get, I always seem to learn something I never knew before, and as Bruno laid out the tale of the Burning Man event, I became more and more intrigued by the minute.

The event had just been held and was attended by roughly 25,000 people. He went on to tell me how these people make the journey to the Black Rock Desert for one week out of the year and apparently become some part of what he called an experimental community, challenging themselves to things not normally encountered in one's day-to-day life. I learned that the theme of the event is to build "the man," which is an enormous structure depicting a man, and that there are a multitude of artists who attend the event. There are striking sculptures, installations, performances, theme camps, and costumes, and all who attend actually erect what becomes known as "Black Rock City." He told me how they perform the burn on Saturday night, using very sophisticated pyrotechnics and the like to ignite the huge structure on fire.

After hearing this, I took Bruno's word that this LLC office was equipped with all the amenities of maintaining contact with the outside world from this very remote location, so we headed on over. My curiosity was piqued, and I was anxious to see this unique headquarters for myself. Danger Ranger greeted us very warmly and graciously, and we began laying out our mission in the desert. He was not only really nice, but, true to Bruno's word, he really went out of his way to help us out. Amazingly, in the short time we were standing around his office getting acquainted and just poking lots of fun at each other, the fax machine rang. He headed over to check it out, and his face lit up immediately. He announced to us that the fax was from the Bureau of Land Management, and it must be for us. We all dashed over to see it, and the whooping and hollering commenced: it was our launch permit! We were all ecstatic as I grabbed the fax and thanked him up and down. We cruised back to our room so I could get to my briefcase, and I immediately called the FAA office in Los Angeles.

"Yes, may I please speak to Mike Coffin?"

When he answered, I told him, "Mike, this is Ky Michaelson of the Space Shot 2000 project. I'm calling to let you know that I've just received what I deem to be our launch permit from the BLM."

"Yes, you're correct. That is your permit. Now, I'm glad you called, because I've got some more good news for you and your team. I've just received the go-ahead to contact Reno, Sacramento, and Salt Lake airports to block out a time where no aircraft will be allowed to fly north within a 40-mile radius of your launch location."

SPACE SHOT 2000

That was the best news I'd heard all day. But, just when things were really looking up, we ran into a snag. His good news was accompanied by some bad news. There was now a condition that I felt certain we couldn't possibly meet.

"However, in order for you to proceed, you will have to prove to the BLM and everyone else that your crew has the capability of maintaining uninterrupted, simultaneous contact with all three of the control centers for the entire duration of the launch."

Needless to say, we had no advance warning of this condition and weren't prepared in the least to accommodate this request. The only phones we had with us were our cellular phones, none of which could even pick up a signal from out in the desert. This little hitch could literally bring the whole project to a halt.

"Okay," I replied. "We understand. I'll get back to you," and I hung up the phone. We needed help, and we needed it fast, so we headed back to Danger Ranger's office to see if he knew of anyone who had a satellite phone. We figured if anyone would have a connection, it would definitely be him.

We cruised in, told him the situation, and he very calmly told us, "You guys are really in luck today. You see, part of the condition of holding the Burning Man out here is that we are required to dismantle everything we build, and any waste made or objects consumed must leave. There's still a very large group of volunteers on site, returning Black Rock Desert to its pristine condition, and they'll have what you need." So he pointed me in the direction of where the group would be.

He gave us the directions, and as we headed out the door, he laughingly said, "Don't worry about what you may see when you get there; you *will* have access to a satellite phone once you hook up with these guys."

I wasn't sure what in the world he meant by such a statement, but we thanked him and drove off.

"Bizarre" doesn't even begin to describe the scene before us as we reached the area where this fine group of people was bustling around, picking up debris. Having met countless types of people from all walks of life, I found myself absolutely dumbfounded at how to react. In all honesty, for the first time in my life, I was nearly speechless. The cleanup crew was not your ordinary, run-of-the-mill group of people. There they were, busy as bees, and completely buck naked—men and women alike. Naked as naked could be! We all sat in the truck and, out of pure shock, couldn't help but giggle. Of course, my wife and the gang immediately nominated *me* to go do all the talking. I just sat there for a few minutes trying to come to terms with how I would even approach them and arguing that perhaps I really wasn't the best person for this particular task. It was all to no avail. I wasn't getting out of it, so I hopped out of the truck and headed toward them. I took a few steps and turned back real quick, only to see Jodi and the guys cracking up at my expense from their little safety zone there in the truck.

"Well," I told myself, "this is for a seriously good cause. I *have* to get my hands on a satellite phone." So, onward I went. A few of the folks stopped working and watched me approach. I called out my very nervous, "hello," and a guy headed to meet me. Of course, I extended my hand immediately but didn't dare look up or down. I made the most direct eye contact I think I've ever made, without so much as blinking an eye, and as we shook hands, I realized I was much more uncomfortable than he was, or anyone else there for that matter. I made my very quick introduction and told him that Danger Ranger had sent me. I'm actually quite sure he thought I was the strange one as I very nervously rambled on about my purpose for coming there and told him I'd be launching a rocket into space in the next few days. I conveyed to him the new government stipulation my team and I were faced with, and that it would really be helpful if I could use their satellite phone system to communicate with the three aircraft control centers. He responded that he and the group would be more than willing to assist us, and that it would be no problem at all for us to use their equipment. He was actually quite excited about the whole prospect and asked for directions to the site. He told me they'd be over just as soon as they were finished with their job and would get everything set up for us.

I only hesitated for a brief moment, and, as if he read my mind, he jokingly told me, "Oh, don't worry. We'll have our clothes on when we get there."

I know my face must have turned as red as a beet, but I just laughed and gave him the directions. As you can probably guess, I got nothing but serious ribbing when I returned to the truck. The jokes were just flying, but I didn't care. I had gotten us the equipment we needed and met some really nice folks at the same time. We headed over to the site, and in no time at all, they all arrived, fully dressed, and got busy setting up all their communication equipment. Suddenly, we had new additions to our team who became instantly as enthusiastic as we were. It was actually really cool. With the tower, rocket, and communication equipment all in place, we spent the rest of the day rehearsing for the launch, which was scheduled for the following day.

SPACE SHOT 2000

Ky stands with the Space Shot 2000 rocket.

I returned the call to Mike and let him know everything was in place. We were all set. Sleep didn't come easily that night, and my nerves were a complete mess. My eyelids were wide open before the sun even thought about rising, and we were up and at 'em at the crack of dawn. We all met at the site to see the sun rise in the desert, and I was extremely apprehensive about the day's events. I couldn't help but think of all the scenarios and ways for things to go wrong. I kept hearing the old saying, "If something *can* go wrong, it *will* go wrong." One of the main reasons I had for fearing the worst was knowing that when the rocket motor shut off, there was a possibility the nose cone could come off at the wrong time. Due to the tremendous amount of air drag once the motor in a rocket shuts off, it's equivalent to slamming on the brakes, and everything tries to move forward. While the nose cone is definitely designed to come off, it can't do so until the parachute deploys, which occurs after apogee. I was just hoping our cone would stay intact until the precise moment, but I didn't really share my concerns with anyone else.

The sun came up, and we were blessed with a picture-perfect day and excellent launch conditions—very sunny, and the sky was clear as could be. We were all pitching in and making our final preparations when the phone rang and startled me. It was the FAA telling me they were ready for us to begin our pre-launch procedures. We had to kick it into high gear, and we wasted no time. Everyone moved *very* quickly, and within minutes, we launched our two weather balloons, each one to a prescribed altitude. We sent the first one up to 1,000 feet and the second one to 5,000 feet. As soon as each one reached its mark, the balloons were cut away by a pyro device. Our wind-speed recording equipment came back to us via parachute, and the readings were loaded onto our laptop computer, where our program then calculated the azimuth and elevation we needed to adjust our launch tower to. Getting the readings and making the adjustments was really eating into our prescribed one-hour timeline.

We only had about 20 minutes left to go on our waiver, when suddenly we saw a truck driving across the desert at a high rate of speed, heading down range. I began to feel like Dorothy in the witch's dungeon, watching the hourglass measuring the passage of time. We obviously couldn't launch until this truck was off the course, so the clock ticked. By the time the thing got out of our way, we had a mere 10 minutes left. Besides our videographer, Jerry and I were the only two standing 1,000 feet from the rocket. Everyone else was *way* back. I announced, "The sky is clear. The range is clear." That signals countdown, so we began from T-minus 60. My entire body trembled with excitement. In just a few brief moments, my lifelong dream was about to come true. When we counted to one, both Jerry and I pushed the launch button together, and it was the most awesome feeling ever imaginable. The second we released it, my eyes filled with tears of pure happiness and joy. This was it, man. This was what I had waited my entire life to do, and here I was, and there it was.

I was instantly mesmerized as I watched an absolutely *huge* ball of fire come out from underneath the rocket. The thrust of the blast caused massive clouds of desert sand to billow out from underneath the mighty rocket motor, and the sound of blast-off was nearly deafening. As if waving it on, the American flag flew high and proud atop our launch tower, as the rocket passed it and headed straight into the sky with easy clearance. The enormous white jet stream against the blue sky was a spectacle to behold, and as I stood there in awe, I very loudly began screaming, "Space, space, space!"

The rocket was over 40,000 feet up when the motor shut off after all the propellant was burnt. Things were looking great, when suddenly the unimaginable happened. From extremely high up in the sky, the rocket completely disintegrated. I instantly feared that my premonition about the likelihood of the cone coming off early had just come true. There was no time for speculation, though, because just then I heard our range safety officer, Bruce Lee, hollering, "Heads up. The rocket has come apart!" In an instant, everything went from jubilation to stone-cold silence as we all waited to hear the rocket pass through the speed of sound. When things go right, you can always hear a sonic boom before the rocket hits the ground. Since we heard nothing, I was very perplexed as to what exactly had gone wrong and why. I knew one thing for sure, though, and that is what goes up must come down, so we kept watching the sky for signs of falling debris.

As I was standing there staring up at the heavens, reality hit me like a ton of bricks. I stood there in total disbelief. We had all worked much too hard

SPACE SHOT 2000

to have this happen to us. My lovely wife, Jodi, saw me standing there, and my face must have told the story, because she very quietly walked over to me and gave me a hug. The second she did, I lost control. I had trapped so many emotions inside of me for two years prior to this day that they all came out at once. I broke down and cried on her shoulder and told her how very sorry I was for having spent so much time working toward this goal, and for spending so much time away from my family, and for spending a lot of my personal savings. I couldn't stop crying; I felt as if I had let everyone down, especially my 92-year-old mother. I had wanted to accomplish this goal while she was still on this Earth so she would be proud of her son. She knew better than anyone what a difficult time I had had in school, and more than anyone how much this venture had meant to me. I longed for my father at that moment, too, more than I had ever longed for him before. I wanted him there so badly, standing next to me, to ease the pain that overtook me at this failed attempt. As Jodi tried to comfort me, I suddenly heard the voice of one of our videographers over my radio. He said, "Ky, I just saw a piece of metal when I was looking through my camera lens."

I composed myself, quickly grabbed the radio, and asked him, "Where are you at?"

He replied, "I'm about five miles behind you."

I told him, "That's impossible."

He said, "Ky, believe me. I know what I saw."

I had been sure the wreckage was in the opposite direction, but at that point, I realized I could very easily have been wrong. I immediately asked the people from the BLM for permission to go into that area. They said we could as long as we followed them. No problem. About 20 of us spread out and slowly walked through the area. I was suddenly overjoyed when I heard one of the searchers holler, "Hey, I found a silver coin!" I couldn't believe we'd find anything, because I truly thought the rocket would be buried in the ground, never to be seen again, and certainly not to be found. A short time later, more good news came. We actually found one of the fins. No sooner had we found that, when another call came in informing us a few of the other searchers had found the nose cone and the booster. This was like participating in a treasure hunt. We searched the whole area very thoroughly one last time and ended up locating every single piece of the rocket, except one fin. We gathered up all the wreckage and headed back to Bruno's, where we disassembled the electronics compartment, and recovered our flight recorders. Much to our surprise, they were still in really good shape. We were able to download information on the flight, which led us to discover why our rocket had destroyed itself. It had reached a record speed of 3,205 miles per hour when the motor shut off. That's when we believe a catastrophic severe wind shear occurred, tearing off the other fin and rendering the rocket unstable. Its instability led to a chain of events causing its mass destruction. In order to back up our theory, we loaded up the video, and while watching it in super slow motion, we soon discovered that the way the rocket disintegrated was kind of a blessing in disguise. Its components flat-spun apart, like a Frisbee, so everything hit the ground at a significantly slower speed, which is why we never heard the sonic boom. Had the rocket barreled to the desert floor at supersonic speed, everything would have been completely destroyed, and there would have been no way to research and learn from this data.

In line with my firm belief that there's always some good to go with the bad, on September 29, 2000, our Civilian Space eXploration Team proved that theory. The Space Shot 2000 was not a complete failure and will still go down in history for a number of reasons. Even though our rocket blew up, we managed to accomplish many things that others had not.

- We were the first civilians to get fully licensed to launch a rocket over 390,000 feet.
- We managed to set a new speed record of 3,205 miles per hour—mach 4.9—breaking the old record by about 150 miles per hour.
- We were able to build a rocket for 1/10th the cost the government would have incurred, because we persevered and never gave up.
- We were the first amateurs to develop software and hardware that conducted wind weighting for a ground-based space launch.
- We established real-time communications with three air traffic FAA radar centers for local aircraft surveillance and launch abort capabilities.
- We proved to our peers that you can get fully licensed and launch a space shot from the Black Rock Desert.

Hopefully, these accomplishments have helped open the doors for others to follow this history-making event. I am very proud of what our team accomplished, but more important, and what means more to me than anything else, is that through this project, I met Jerry, who has now become my lifelong friend. Like I said, there's always good that comes with the bad! You just have to be able to recognize it and appreciate it.

Chapter 20

Rocket Withdrawal

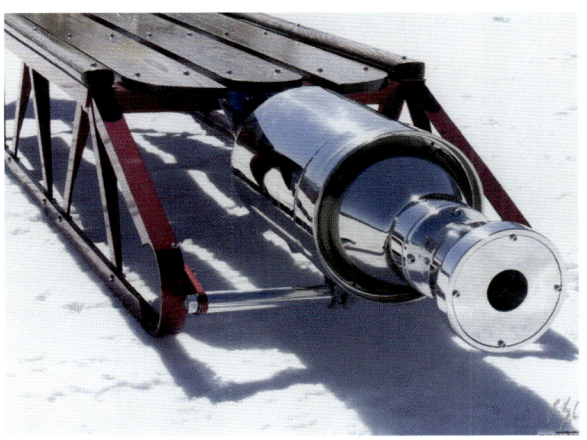

A JATO-powered sleigh could put Santa out of business.

Jerry and I were more determined than ever to make another go at getting a rocket into space, but in order to wind down from the bureaucratic nightmare we had just endured, we decided to wait until at least the first of the year before beginning our design work. Besides that, we both really wanted and needed to spend more time with our families. With both feet planted firmly on the ground (except when they were planted firmly in my reclining chair), I relished my family time and was able to slowly release all the stress and tension of our failed attempt in a very positive manner. Life was truly good again, and I was quickly reminded of all the blessings bestowed upon me from the quiet comfort of my home.

But during the first 30 days after the launch, I think I went through some sort of rocket withdrawal. If there were a pill for that, I would have taken it, because it was truly painful. It seemed like every time I closed my eyes, all I'd see were rockets. The fact that every time I walked into my shop I'd have to pass by all my rocket toys really wasn't the prescription I needed and wasn't helping my withdrawal one little bit. I stood out there one day, looking at my rocket-powered car, motorcycle, snowmobile, go kart, wheelchair, and backpack, and started reminiscing back to my boyhood, to one of my favorite things: my sled. Now, you just take a guess at what happened next.

Well, it just so happened that I had a JATO rocket that needed a purpose. I stood there and thought about the big thrill and controversy I could cause with just this one rocket. That was it. I set to building my rocket-powered sled, and if I hurried, I could get it done without Jodi even missing me. After all, I was supposed to be taking a break from rocketry, right? Wrong!

It's like an addiction for me. I worked in the shop when I could, and before long, my new creation was ready to rock and roll. Winter was upon us, and I finished it just in time for our Super Bowl party. Jodi and I host this annual event and have people come from as far away as Sweden, so I'm always trying to come up with new and innovative ways to entertain our guests during their visit to our frozen tundra. I had my agenda now; I was breaking out the sled. I made sure to invite some members of our local police department to the party, because I planned on debuting my new JATO sled, and I didn't want any problems. The street in front of my house is a four-lane thoroughfare with lots and lots of traffic, and the city keeps it plowed really well. It was the perfect launching pad for this rocket sled. I tested it out a few days before the party, during the day when traffic was at a minimum, and it worked great. I told the local law enforcement that I had plans to give some of the party guests a chance to jet down the roadway, so they were nice enough to actually block off a lane for me. Jodi and I gave rides to people throughout the game, and the police were actually very impressed when they saw the safe manner in which we did so. Everyone really enjoyed watching all the fire and smoke shoot out the back as it propelled down the freshly plowed lane. As is usually the case whenever the mad scientist opens his shop door, we managed to stop a lot of traffic, and a lot of people actually parked their cars at the public parking lot across the road just to watch us. It was a great time and managed to pacify me just long enough before my focus shifted to the big task at hand.

Chapter 21

The Day the World Stopped Turning

Jerry finished up the design and blueprints for our Space Shot 2001 rocket, and everything was going even better than planned. We got it built in record time and actually received an invitation from the GSTA to display our newly designed rocket, along with my entire collection of rocket-powered vehicles, at the 45th Annual Rod and Custom Show. I was standing next to it when a guy came over and introduced himself to me. The man, Eric Knight, immediately went on to ask what it would cost him to have his company's logo printed on its side and if that was something we'd even consider. I have to admit that none of us had ever thought seriously about having a potential sponsor, so I told him I'd have to let him know. In the short time we spoke, I really liked him, and we ended up getting together shortly after the show. That's when I had the opportunity to meet his family, who were all just as genuine as he was, and it took me no time to realize that he could become a very significant asset to our project. With Jerry's blessing, we named our new rocket the *Primera*, and our new project name became the *Primera* Space Shot 2001. I was also given the chance to do a number of local TV and radio talk shows to promote both the car show and the Space Shot 2001 project, so we got an abundance of media exposure for our Space Shot rocket.

Eric Knight joined the team and volunteered to work on the avionics, along with a team of other rocket and ham radio enthusiasts he recruited from the East Coast. Our new rocket was about 4 feet taller than its predecessor and carried nearly 75 pounds more propellant. Not only was this rocket more powerful, it was also more sophisticated, with a major change sure to add more stability. Short, canted fins rapidly spin the rocket (six times per second), and its angular momentum helps it resist sideways forces. We canted the fins on this one, so it would spin like a bullet, and made several other modifications to the fin area that would guarantee—no matter how much fin flutter or side load we encountered—the fins wouldn't tear off. With the help of Eric, the thing was literally loaded to the gills with very sophisticated electronics and avionics. I'm talking two onboard pinger tracking transponders, GPS receivers, a telemetry transmitter, onboard flight data recorders, a three-axis accelerometer, a three-axis magnetometer, and television transmission to allow us to see black sky and the curvature of the Earth via our laptop computers. The sheer sophistication of this project was truly the most exciting part of the whole thing for me, and it took us the entire summer to make all the preparations for our scheduled September flight. We were certain that our little group of folks would make history; we couldn't have known that history would be made through the incredibly tragic events that took place on September 11, 2001, the day the world was forever changed.

Like so many other Americans, I watched in sheer horror as our beloved country was attacked when four U.S. airplanes hijacked by spineless terrorists crashed into the World Trade Center, the Pentagon, and a field in Pennsylvania, killing more than 3,000 people in a matter of hours. The entire world shifted gears, and because of that horrendous act of terrorism, all aircraft in this country was grounded as a means of prevention of further massacre and turmoil. Suddenly, our mission seemed minuscule in the shadow of all the tragedy, and everything was put on hold indefinitely.

In the aftermath of September 11, we pressed on and continued working on our television show. Most of the present-day footage had been shot, so we began reenacting some of my childhood follies. Jerry's son, Tyler, was handed the job of portraying me as a young boy, and the production company began shooting scenes depicting the good old days when my friends and I went diving with my homemade diving helmet. All the tragedy and despair going on around us had made me realize how very precious life really is. As I watched this fine young man and this group of young kids pulling a little red wagon with the diving helmet in it around Lake Nokomis on their way to the dock that was just as I remembered it, I was moved to tears. I was saddened to think that today's children will never truly know what it was to grow up without a care in the world. There they were, one of the boys carrying a fishing pole, and the other kids pushing each other around, just acting like kids, and that had been me, back in a time when the word "terrorist" never even existed. This production was yet another wholesome distraction, and I soon found myself pouring my heart and soul into every scene. I especially had fun with the reenactment of me trying to fly on my bicycle with an ironing board strapped to it. I had told Tyler, "Remember to keep pedaling, 'cuz you're really trying to fly." Tyler pedaled like mad down the hill, while the other kids ran alongside the bike, until they fell and tumbled downhill, head over heels. I really *had* tried to fly back then; watching him in action brought me back to that day and made me realize these sorts of memories are what make life worthwhile. Stepping behind the camera and seeing the pure innocence of those kids depicted, believing what they were doing would actually work, just made me laugh out loud.

After we finished filming those scenes, we decided it would be fun to look up some of my actual childhood friends and interview them. What an enlightening experience that turned out to be! I always thought I was just one of the boys, but they all claim to have known there was something special about me. I was very humbled by their comments. I knew I couldn't possibly complete this film without some input from the one person who knew me the very best of all, so we proceeded to interview my mother, who was 92 years old at the time. The camera rolled and so did my tears as I sat and listened to my mom tell me how very proud she was of me and that she really believed I would be successful at launching my rocket into space. She finished the interview by telling me she just hoped and prayed she'd live long enough to see me do that. Her faith in me is one of the main reasons I've persevered in my goals and continue to do so today.

With the film in the can and winter upon us, I needed to keep myself busy until our next window of opportunity for a launch approached. So I decided to build a twin-engine motorcycle identical to the one I raced back in 1960. I bought two 1959 BSAs, extended the frame of one of them, and totally rebuilt the engines. I put every ounce of my being into this project and gave it all the love I knew how, and if you sit back and look at it, you'll see what I mean. To this day, I don't think I'll ever even fire it up, because if I do, the pipes will turn blue, oil will get all over it, and I think it's too beautiful to take that chance. I'm planning on having my own museum some day, where I will proudly have this bike on display. I've publicly displayed it once already, at the GSTA Rod and Custom Show, and am happy to say it brought me a first-prize trophy. My good friend Bill mused about that as he told me, "Ky, if you stop to think about it, the worst prize you ever took in a car show was first."

Chapter 22

Mother Michaelson

By March 2002, things really started picking up again. Jerry started working with all the necessary federal agencies, and by the first of June, we received our permits and all the licensing required for launch later that month. Things really fell into place quickly, unlike our previous attempt, where we had to jump through so many hoops.

Sadly, though, my personal life was in complete shambles. I was an emotional mess, because I found myself dividing all my time and energy between two things that meant more to me than the world. As our launch preparations were falling into place, my mother, who was then 93 years old, lay in a hospital bed, recovering from colon cancer surgery. As if that wasn't hard enough on her, the doctor had instructed the nurses to have her sit up in bed for some exercise, and when they attempted to follow those orders, her body was so weak that she then suffered a heart attack and a stroke. I did nothing but hope and pray that my mother would recover from these very serious problems to see me succeed in what I had worked so hard for over the past seven years. I was in very close contact with the hospital, doctors, and staff, and we decided to move forward with the procedure of removing a very small balloon they had placed in her heart. They told me she had less than a 50 percent chance for survival. I spent as much time with her as was humanly possible throughout her ordeal. I would sit at her bedside, place my hand over hers, and just reflect. I thanked the good Lord for blessing me with her and allowing me to have her here on this Earth as long as He had.

As I looked at her, I couldn't help but think that she was, without question, the youngest 93-year-old woman I had ever known, filled with more spit and vinegar and life than most of my much younger acquaintances, including my own children. I still can't think of another human being who ever taught me so much throughout my own life. I sat and talked with her about all the good times my siblings and I had had as kids, how we were always warm and never went hungry, and how loved we had always felt because of her and my dad. I even reminded her how I had put snakes in her garden in my attempt at getting out of working there. She laughed and told me she remembered. I really got her going, too, when I reminded her of the chicken soup she'd made out of my pet chicken. I said, "Yeah, well, if you think *that's* funny, Mom, the sandwich you just ate here at the hospital was actually made with your beloved cat." We all got a good laugh out of that one.

She slowly began to mend and made a request that we were all more than happy to oblige. She told us she felt her time was near, and that she wanted to spend her remaining days at her home, where she could be with her husband, her cat, and enjoy her beautiful flower garden. It was June, and everything was in full bloom, and the weather was nice and warm for her. The hospital arranged to have a special bed brought in and a hospice worker to help take care of her. Throughout this time, our launch date was also upon us, and I was absolutely torn. In a way only my mother could, she somehow persuaded me to carry on with the launch and told me she'd be watching. I stopped by her house on Sunday, because we were heading to Black Rock on Monday. In some ways, I almost wished I hadn't done so, because it was a shock for me to see her in the condition she was in. She was gasping for air, and she told me she knew she was dying. Hearing my mother speak those words to me was almost more than I could bear. As difficult as it was for me to see her so weak and in so much pain, I stayed with her for the day. When it came time for me to leave, I told her how much I loved her, and she acknowledged she knew, that she loved me too, and she wished me good luck as she bade me farewell. Walking out her door that day was the hardest thing I had ever done, and my mother was very heavy on my mind as I boarded the plane and headed to Nevada the next day, not sure if I would see her beautiful face ever again.

ROCKETMAN

Ky with his mother, sister Maradee (left), and sister Marlee.

Ky with his mother, daughter Miracle, and son Buddy.

Chapter 23

Mother Nature's Wrath

The AST office had actually been more generous this time around and had granted us our launch license a full week in advance. They'd also gone ahead and added several new provisions that would definitely make things a bit more difficult but certainly not impossible. With the license in my hand and my mother's good wishes and blessings in my heart, I felt good things would come, and we'd get through the newest challenges they'd presented: making certain there were now no trains within 30 minutes of our launch, adding Fallon Naval Air Station to our list of air controllers requiring constant contact, and shortening the launch window and the allowable winds that we could launch under. Like I said, more difficult but by no means impossible. The whole crew was out in the desert when Jodi and I arrived. This was our first chance to officially meet all the new members of the CSXT Team, and it was an incredible sight seeing so many people working so hard to make this new space shot a reality.

For some reason, it seems everything I have ever worked for has just never come easily, and this project was no different. I first realized this when I noticed there was obviously something bothering some of the team members. When I asked one of them what was wrong, he told me we were missing a crucial part of the rocket but that they had already made a call back to Minnesota to have a new part machined. They had been told it would be flown in the next day, which really wasn't very reassuring given the already *very* tight schedule we were on. This would push us back by a full day. Well, that wasn't good news by any means, but we gathered our composure and decided to make the best of it; we'd get past it and move on. No sooner had we attempted to lift the spirits of the team than yet another bombshell was dropped. The electronics team discovered that for some reason one of the computers was prematurely emitting the signal to set off the pyro device designed to cut the rocket in half. That was a very, very serious malfunction, because if that were to happen, it would not only create an extremely dangerous situation but would also destroy any hope for a successful launch. The team worked well into the night in an attempt to correct the problem, but to no avail. They ended up having to actually redesign part of the system to ensure the premature detonation wouldn't occur. With heavy eyelids but a renewed window of hope, we all headed to town for some much needed rest for the big day.

While most of us headed back out to the desert the next morning, one of the crew members drove back to Reno to await the delivery from Minnesota. Little did we know at the time how lucky that volunteer was. The Black Rock Desert is a very beautiful area, but its weather is totally unpredictable. It is a place where anything can happen, and if you aren't fully prepared, it can actually become deadly. I say this now, because I learned it that very day. It was about 3:00 in the afternoon when Mother Nature decided to show us who was boss, and we were, without question, totally unprepared. While we were all working hard to get the rocket ready, none of us bothered to ever once look to the horizon, or we would have had time to batten down the hatches. The winds snuck up on us, and when I say deadly, I mean they were deadly. I have never in my life experienced a sand storm, and the one we found ourselves smack dab in the middle of was even more powerful than anything Hollywood could muster. We clocked the winds at over 50 miles per hour, but I have no doubt that with the gusts, they were more like 80 miles per hour. The sting and pain of the sand pummeling against our bare skin was more intense than any BB gun I had ever been hit with, and we had no time to seek shelter or cover ourselves. We had enormously expensive camera and electrical equipment that we needed to try to pack away, and in minutes we found ourselves all scrambling to save our launch site. The small tents some of the crew members had slept in were blowing aimlessly and ferociously across the desert, completely gone from sight in seconds flat. We had a larger 40-foot tent for our base, and we all clung to it for dear life. We didn't know if we would be able to save it or not, but we held on with all our might, and Mother Nature blew those winds at us for over an hour.

MOTHER NATURE'S WRATH

As suddenly as they came upon us, they subsided, and the desert that had minutes before sounded like a freight train became eerily quiet. It was one of the most frightening, yet strangely exciting, experiences I have ever been through. After we finished giving praise to the Lord for keeping us in one piece, we spent the next eight hours checking and cleaning the equipment. It had been blasted with sand and was in total disarray. Even though we had lost the big battle, we somehow won the war—we saved almost all of it. Our volunteer crew member came back to our site in total awe and disbelief at the shambles before him. As he handed over our special delivery, we were quick to tell him how lucky he had been to head into Reno earlier that day. We packed up what was left of our camping gear and prepared to bunk cozily cramped together back at Bruno's; there would be *no* camping out on the desert that night.

Morning came, and we all headed to the restaurant for Bruno's wonderfully famous hefty breakfasts. Talk about windburns! As we all gathered, we found ourselves comparing what we were certain would be scars from the sand that had felt like golf balls pelting our skin the day before. Some of us looked like we had road rash, and it stung like the dickens. Anticipating another long, hot day in the desert, we ate as much as we could so it would last us a long time. The burning, dry desert heat is so incredible that you can dehydrate at a very rapid pace, so when you're out there, you consume mostly fluids throughout the day. Almost every year, you hear stories of people who make the mistake of going out there without adequate water supplies, and they usually end up with either serious medical problems or sometimes succumb to death. Since we were still licking our wounds and had *no* intention of becoming one of those statistics, we were more than prepared. With full stomachs and gads of water, we bade our farewell to Bruno and headed back out. With all the obstacles we still had to cross, as well as the setbacks and new members without much experience, we had many new operations to perform before we could launch. This would be our rehearsal day, to ensure things would run smoothly when the time came. That was the plan anyway . . . new day, plenty of rehearsal time; no problem, right?

We arrived on site and spent a couple good hours setting up what had been blown away the day before, but today things were different because *today* I had open sores on my legs, and they were stinging from the heat and the sweat. This prompted me to take a moment's break from what I was doing and actually look toward the horizon. Lo and behold, Mother Nature was brewing another storm, and I could clearly see the front moving quickly

in our direction. "Hey, guys," I yelled. "Break it down; we have to break down camp *now*. Look what's coming!" Everyone stopped and stared, and in no time at all, we all scrambled to undo what we had just done before the winds tore it apart for us. We got it done in the nick of time. I can't even begin to tell you how disappointed we all were, but even more than that, sheer exhaustion had set in. We were every kind of exhausted you could think of, and we just felt plain old beaten. None of us had anticipated having to deal with forces we couldn't control, and it literally took the wind out of our sail. We headed back to Bruno's to call a meeting to discuss our next move. Two of our five launch days had already been used. The AST wanted an update.

My old pal, Randy, and another AST member named Mark sat in on our meeting. My personal frustration with this whole endeavor had now come to the surface, and I found I could no longer contain myself. Forces of nature are one thing, but the AST was another. I just felt they were truly doing all they could, without openly stopping us, to discourage us from succeeding. Our production crew was filming our meeting, because we capture it all, and the entire crew was present. What better time for me to explain my displeasure and frustrations with the AST than now? I reiterated to them that they had never once so much as encouraged us in any way, and that all we seemed to get from their agency were multiple roadblocks, red tape, and lip service. Randy sat back in his chair and very callously replied, "Well? We have a motto in our office that goes like this: 'We're not happy until you're *unhappy*!'"

I quickly retorted, "I believe you, Randy. I totally believe that's true. The fact of the matter is you *don't* have our best interests in mind, and for what reasons, I have no idea. We are not a bunch of radicals. We love this country, and we obey its state and federal laws. All you keep saying is how you're here to protect the public." He glared at me, but I continued, "It is perfectly legal to launch a rocket to 100,000 feet. All you need to do is file a waiver with the FAA. The FAA could care less how many people we have in the launch area. Yet you guys, the AST, have regulated us to the point where we can't even have one person down range or our license is invalid and we face a $100,000 fine."

Randy sat straight up, leaned forward, and looked as if he were about to explode, but I wasn't giving him a chance to speak. He was going to listen to *me* for a change, 'cuz I had heard enough of his crap. "I've said this before, and I'll say it again," I continued. "Aircraft are allowed to perform stunts over the heads of a hundred thousand people at air shows, and each and every year people are injured or killed, but the FAA hasn't taken it upon themselves to regulate those events to the point where it's nearly impossible for them to perform. That, Randy, is exactly what you've done to us, though. I can't believe you have the audacity to step in with all your bureaucratic red tape when you don't even have enough sense to come up with an official application for a license to launch." My last remark to Randy, whose face was now an angry rosy red, was, "You know, I suppose your office has actually given you some sort of an award, or high grade, for your evasive performance. You *must* be proud, Randy."

My entire crew sat in stunned silence as the cameras followed Randy and Mark, who got up angrily and made their hasty exit through the door, stating they'd heard enough. The moment they left the room, we felt rejuvenated, and we immediately came to a rather risky conclusion. No one ever said we couldn't attempt a night launch, now did they? New plan. We moved, and I mean quick! Randy and Mark were waiting in the lobby for our meeting to adjourn, so we informed them of our new plan, and they insisted we contact the railroad to get the time when the trains would be in our launch area. The railroad told us because of the possibilities of terrorist activity, it could not reveal that information to us. Just like that—another roadblock we couldn't handle by ourselves. Not to fear, though, because luck was finally on our side. A couple of our newest team members were with the BLM, and they called and got the information for us. If it hadn't been for them, we would have been beaten right then and there. We felt a renewed sense of determination as we kicked it into high gear and headed back out to the desert, with Randy and Mark in hot pursuit.

We wasted no time in making our preparations. We readied the rocket and got it into the launcher for its short flight into space. Unfortunately, though, the weather didn't appear to want to cooperate with us at night either. The heavens were cloudy with patches of blue sky, and the winds were intermediate. We were all under a tremendous amount of pressure, but after all the bull with the AST, we decided to take a chance and go for it, hoping the skies would miraculously open up and the winds would calm down. Jodi and the crew started filling the weather balloons with helium, while I made the calls to the air controllers in Salt Lake, Seattle, and Oakland. I got through to the three with no problem, but when I made the call to our newest additional center at the Fallon Naval Air Station, right there in Nevada, all I got was an answering machine. I couldn't believe it! I frantically tried it again and again, but there was no answer. The pressure was too much for me to handle. Everything was going wrong, and I simply had no choice but to stop the launch. It just seemed like everything was against us.

I immediately turned to Mark and Randy and said, "You guys ought to be ashamed of yourselves. The Fallon Naval Base is nothing more than a catch-22.

MOTHER NATURE'S WRATH

Even if everything was *perfect*, we wouldn't be able to launch because we are unable to reach all four control centers." I called the team together and told them, "Look, we still have two more launch days, guys. All we're doing here is beating our heads against the wall. Let's head back to town and regroup. We need to try to get AST to loosen up a little bit so we can achieve our mission." We deflated the helium balloons and began packing up. Randy and Mark took off, and I'm certain they shared a little victory celebration on their own. Meanwhile, we all headed back to Bruno's to try to get some sleep.

With the dawning of the next morning, I was delivered the news I had hoped I would never live to hear. My dear mother had just passed away. That was the final blow for me. I was completely and utterly devastated. I called Jerry to my room, and my face and my tears told the tale. I let him know if he and the rest of the crew wanted to launch the rocket, they should stay, but I needed to go.

Jerry hugged me for a few minutes and very calmly said, "No way, Ky. We're not going any further without you."

I broke down right then and there, and just cried out, "Do you see? If only Friday's launch attempt had been successful, I would have been able to keep my promise to my mother that I would get our rocket into space while she was still on this Earth." I let it all out, every bit of it. All the hurt, all the anger, and all the disappointment spewed from within me, and my body just shook with raw emotion. Jerry left Jodi and me to be alone, and she comforted me as best she could. I released my feelings until exhaustion set in, and my body sank into the couch, where I sat motionless as Jodi left me to go break the news to the rest of the crew. Loneliness enveloped me as I reflected back to my last visit with my mom just a few short days earlier. I knew she was in a much better place now, resting in peace, and then a thought came to me. I told myself that when we *did* finally launch our rocket into space, she'd look down on her son from her front-row seat, and I'd hear her say, "Ky, I knew you could do it."

As the news of my mother's passing spread, it was unanimously decided the launch attempt would be aborted. I could not go on, and my crew just didn't have the heart to go on without me, knowing what I was going through. I was ever so grateful for all their caring and understanding, and with the heaviest of hearts, Jodi and I boarded the plane to go see Mother one last time.

Chapter 24

Primera Space Shot 2002

In all honesty, those first few months after Mother's passing, I tried everything within my power to walk away from ever building another space shot rocket again. I felt as though I would never overcome the unbearable grief and longing for my mother that had totally consumed me, and I truly wasn't myself for a long, long time afterward. Jerry and the rest of the team members kept in constant contact and supported me throughout the worst point in my life. Time passed, and my heart started mending, and they convinced me that we would try again—that it's what my mother would have wanted me to do. Their words inspired and comforted me, and as hard as it was, I knew they were right.

Ky and Buddy outside their Minnesota home prior to Space Shot 2002.

Jerry took it upon himself to once again get things back on track and called Washington to see if it was possible to recycle our license for September. They said they'd accommodate us as long as we didn't make any changes to our rocket, but our license now came with a clear set of exceptions and several changes to procedures we'd followed in the past. We now had to scan nearly 200 square miles of the desert playa in search of campers or other people, provide a fail-safe plan to secure all possible entries, obtain the schedule of all trains passing through the area, and perform a thorough aerial surveillance prior to launch.

I wasn't sure where to begin with trying to literally close off such a large area of public property, so I called the Bureau of Land Management and asked. They told me it could be done and gave me the procedures to follow. The first step was for our team to run an ad in the local newspapers for 30 days notifying the public of our intentions. If anyone complained, a special hearing would follow to review our rather unique request. We ran the ads, and I called them back exactly 30 days later. There was good news, and there was bad news. Thankfully, no one had protested, so they gave us a permit to block off the playa. What they hadn't mentioned to me before, though, was that it was necessary for us to present them with a $1 million bond to close the deal—the sort of detail you'd think would merit some special warning. There was no way to come up with a million dollars on short notice, and I wasn't going to give up. I argued that there was nothing out there to kill or ruin, and so they dropped the bond demand, and we settled on a million-dollar liability insurance policy.

Finding someone with a plane who could fly over Black Rock and scout for anyone who'd maybe been camped there prior to our postings proved to be no easy task either. The closest commercial airport was in Reno, which was over 100 miles away. The plane would literally have to be at our beck and call for nearly a week, and after making numerous phone calls, I soon found it would be virtually impossible and very expensive. This requirement was going to take a good deal of ingenuity, so we pushed on.

With the legal right to keep the public out, we had to devise a plan to enforce it. There are six entrances onto the playa, and we were required to have signs posted at each one warning the public of our activities and indicating no admittance was permitted by anyone other than our team and support group during the four-day schedule. The BLM rangers said we could hire them to give us a hand with our security and that they'd have the authority to arrest anyone not willing to obey the rules. Eager to comply, we graciously accepted their offer and knew that making and posting the signs would be no problem.

Obtaining railroad schedules had proven a challenge the last go-round, so I asked Bruce Lee if he'd try calling Union Pacific Railroad's corporate office. I hoped maybe he'd have better luck explaining our mission than I had the last time I called. It turned out he did. He was immediately put in touch with their PR office, and they were intrigued by his request. In no time, they fully understood our predicament and actually had Bruce hold while they patched him through to the dispatcher for the area. (It gets better.) It turns out the dispatcher's son's favorite hobby was rocketry, so she was more than willing to help us, and after a very chatty conversation, they got everything taken care of. It was well worth the pain of listening to Bruce rib me about how to get things done using charm and wit, two characteristics he apparently had that I did not. All kidding aside, he came through for us when we needed him most, and things kept moving.

Eric Knight, our avionics specialist, made a discovery that also literally saved the day, and he quickly became my second unsung hero. He happened to be looking over a map and spotted a small town called Empire located about 15 miles from Black Rock. As he took a closer look, he noted what appeared to be a very small runway in the map's detail of Empire. In addition to all his avionics expertise, he's also our resident ham radio expert, so he started scanning his address book to try to locate any potential fellow operators in the area. He found one by the name of Ken Samuelson and wasted no time making contact. He learned that the entire town is owned by U.S. Gibson Company, a sheet rock manufacturer, and there was in fact a runway located directly next to its plant. Eric continued the conversation and laid out our mission to Ken, who immediately became very enthusiastic. So much so that Eric took things one step further and asked if he could possibly help us out with communications. He was thrilled, and told him, "No problem." He pressed on and asked if the runway meant someone in town maybe had a plane. Lo and behold, Ken had one, and by the time the conversation ended, we had ourselves an aerial surveillance pilot too. Everything seemed to be falling into place. With a renewed sense of confidence, we submitted all the necessary documentation back to the BLM, all of which clearly demonstrated our ability to meet all criteria. With that task behind us, it was on to the next.

Our recollection of how Mother Nature had totally demoralized and beaten us to the ground back in June prompted us to begin work on the next phase of the project: shelter and a launch control center that could withstand the brutal winds, dust, and intense heat. We used a 28-foot enclosed trailer that resembled the kind that food vendors use, with two huge openings on the sides. We set it up with lights, electricity, a PA system, TV monitors, and even threw in a table and some chairs. It darn near had all the comforts of home, and if the weather turned on us, all we'd have to do is latch and lock the doors, and all of us and our equipment would be safe from the elements.

Just about the time we completed that, our new launch license arrived. Our whole team was totally upbeat, and with everything in place, we felt destined for success this go-round. Our four-day launch window was scheduled for September 16 to 19, 2002, so Jodi and I decided to head down a few days early. I've got a long-time school friend in Nevada by the name of Dennis Kottke, so we decided to pay him and his wife, Diane, a visit. They have their own need for speed and run a very successful racing enterprise called Sierra Sierra, and their team competes in formula auto racing. They also own two Russian jets and told us they'd be racing them at the Reno Air Show while we were there. I wouldn't have missed that for the world, as air shows have always been a huge love of mine.

After a fun night of catching up on old times, we headed out to the show the next day. The weather was great, and they were totally pumped and as anxious about their performance as we were about our upcoming launch. We received VIP treatment and experienced the thrill of watching the high-speed military jets at close range as they roared over the crowds, performing their maneuvers, with their afterburners belching out huge flames of fire and black smoke. Their sheer display of undeniable power and might mystifies me and makes me proud to be an American. I believe that feeling was shared equally amongst the crowd, as we all saluted those brave pilots and applauded their demonstrations. The air racers followed, and there wasn't an empty seat in the house. Air racing is an unbelievably expensive and very, very dangerous sport. Even with all my past stunt work experience, I know there's not a thread of a chance that I would have ever pursued that particular need for speed. Not only do they full-out race, but these guys also do stunt flying and have acrobatic teams that perform in between the races. The sport draws hundreds of thousands of spectators, and the excitement is intense. We were fortunate enough to see Dennis and his race team end up in second place in the points standing and walk away without injury. It was a blast and the perfect way to keep our minds occupied on something other than the launch, at least for a little while.

ROCKETMAN

The time quickly came for Jodi and me to head on down the road toward Black Rock, so with Dennis and Diane's best wishes, we were on our way. The whole crew arrived on Monday, September 16, and it was like a family gathering. The mood was extremely upbeat and optimistic as we all got right to work. A four-day window can go by very quickly if everyone doesn't pull together and stay focused, or if unforeseen problems occur, so we wasted no time. The avionics team changed the batteries in all the equipment while we spent the day assembling the rocket and getting it placed in the launch tower. It was amazing how having good tunes cranking on our new PA system kept everybody moving to the beat; that was a luxury we hadn't had before, and it was very well received by everyone. The guys even had a surprise tune for me. They brought along a copy of Elton John's "Rocket Man," and as it echoed out across the playa, I couldn't help but stop and look around and marvel at how far we'd come. The equipment and the technology this group of people had gathered and put into operation was nothing short of amazing, and I was overwhelmed with pride at everything we'd accomplished as a group. I made a solemn vow to myself right then and there that whatever was about to happen wouldn't be for lack of effort; we had all given 110 percent, and if this attempt failed, I had no more to give and would just walk away. With everything in place and ready to go, we all headed back to Bruno's once again for the usual: food and an attempt at sleep.

Eric Knight gives a final check of the *Primera*'s electronics before installing them in the rocket.

PRIMERA SPACE SHOT 2002

We arrived back at camp at 4:30 Tuesday morning. It was really cold and windy, and the skies were much too cloudy for a morning launch attempt, so we waited to see what would happen as the day went on. The winds changed directions but unfortunately not in our favor, so our evening window was also unattainable. Everyone packed up, but Jodi and I stayed behind. There was something I just needed to do, and they all understood. She and I cruised out to the launch tower, and I got out and just stood there looking up at our rocket. I imagined with all my might that my mother was standing there beside me, telling me how proud she was of what we had accomplished. I spoke to her in silence as I walked up to it and, with a felt-tip marker, wrote "Mom, I love you," hoping she could see me and read my writing.

On Wednesday, there was once again no chance for a morning launch, but things became a bit more promising by afternoon. We decided to go ahead and do a dress rehearsal to walk through all the procedures and iron out any unforeseen bugs. We were launching our weather balloons as the members of the Office of Space Transportation, in particular, Paul Wild, began hovering over us, making sure we met each requirement to a tee. Paul was all over Jerry, asking questions that should have been totally obvious or could have been answered later. Jerry politely asked him to stop talking at such a critical time, as he was disrupting his concentration, which could cause him to make a mistake and endanger the launch. Things were tense enough, as we all knew the slightest deviation would result in a $250,000 fine and five years in federal prison. None of us was real keen on the idea of making some haphazard mistake that would drastically change our lives forever. Jerry continued pleading with Paul, telling him his job was not to interfere and that he had no right to interrupt him during the preparation. That's when Paul's true colors were revealed. He flew off the handle in a very unprofessional manner. His face turned beet red as he yelled, "If you don't like it and you don't answer my questions, I'm going to call Washington, D.C., and stop this launch." Things had been tense enough without the added confrontation. I didn't say a word, because I didn't want to give these guys a way out. From the very start, I had felt it would be a long shot at best that we could even meet the criteria of the launch license, but we had, so I just wasn't sure where this would lead. There were a few more heated exchanges before Jerry, Paul, and Mark had a private meeting. From what I was told, they agreed with Jerry, and all sides decided to go forward.

As seriously cold-hearted and extremely difficult as they were to work with, we had no choice but to play by their rules. The worst part was that their office had told us they were there to facilitate and work for us, but they quickly seemed to forget that. If they had been on my payroll, they would have been fired just for having major attitude problems and for their total lack of people skills. They truly did nothing but discourage us, just as they had done to all the others who had traveled this same road before us.

As all the confrontation and meetings were going on, our friend and aerial surveyor, Ken, and one of the rangers had made a sweep of the desert. They came back with news, and it wasn't good. They had spotted a number of people from the Burning Man festival cleanup crew still out there about 5 miles down range. They'd apparently camped there for about two weeks, and when he approached them and asked them to leave, they released their dogs after him. He decided it was best to leave well enough alone and took off. My stomach just sank, as I had no idea how to handle this new situation. After some careful thought and consideration, we decided to send Mr. Personality, Bruce Lee, over to see if he could once again work his magic.

A small group of us piled into the truck and headed off to find them. Bruce strolled over and introduced himself. They realized we were probably there to have them legally removed because their permit had expired and ours was now in effect, but Bruce very diplomatically defused the situation by inviting them to come watch us make our launch attempt. We all figured these people were a bit on the rebellious side and might display some aggression, but as he continued talking, they became more and more intrigued with what we were trying to do. They were actually very friendly toward him and said it was real thoughtful of him to invite them to watch. As Bruce turned and headed back toward the truck, we could hear them excitedly saying, "Man, this is really, really cool." Just as he climbed back in, about 30 of them jumped into the back of their various trucks and vehicles and were headed off in a cloud of dust racing toward the launch area. They got ahead of us, and I watched in horror as they ran over all our wires between the tower and our control center. I hopped out to check for damage and was amazed that everything was still intact.

It was truly a sight watching all those modern-day hippies jumping out of their vehicles, shouting and screaming, and singing and dancing to the beat of a different drum. Their dark brown skin was roasted from the hot desert sun, and they were covered with powdered dust from head to toe.

PRIMERA SPACE SHOT 2002

They reminded me of the Woodstock generation, and their dancing, body piercings, and overall appearance was a complete shock to all the OST officers and rangers standing around exercising such formal diplomacy. They'd never seen anything like it, and I simply couldn't resist saying, "See? I told you I had all kinds of friends!" It was amazing how completely happy and carefree these people were. I kept wondering if I could possibly be that happy having just camped on the desert with no shower or running water and hair matted to my head. I didn't think so. Their enthusiasm brought us to an even higher level as the day progressed. I couldn't resist asking the younger ones if their mothers knew where they were.

They all laughed, and one of them replied, "Oh yes. My mother even sent me a dress for the occasion!" They roared with laughter at that comment, because that was about as far fetched as you could go. Many of them were nudists, and dresses were definitely not part of their culture or lifestyle. It was all about leather, tattooing, and heavy, heavy body piercing. We did stop and take notice that there were a few of them not wearing any clothes, so we informed them they needed to be dressed to be at our site. They had no problem with that, so while the select few scrambled for clothes, I decided to further blow their minds by demonstrating another form of rocket power. It just so happened that I had my rocket-powered mountain bike along, so I brought it out and fired it up. They all wanted their pictures taken with my crotch rocket, and I became an instant hit. They knew I liked them, and you could definitely feel a mutual respect between our two very different groups of people. They ended up hanging out with us for about two hours until we had to once again scrub the launch due to the high winds. We told them they were welcome to come back for our morning attempt, but they'd have to be there by 4:00 a.m. I figured most of them would still be up at that time anyway, so it wouldn't really matter. Oh, to be young again. As it turned out, a few of them actually decided to stick around and camped out with our guys, which was pretty neat, or so we thought.

Most of us headed back to Bruno's for some of his world-famous ravioli. Not only are the portions huge, but the heartburn that follows is too. A little Pepcid or Tums is a must, but the slight discomfort is well worth it, because there's no better pasta anywhere. Bruno's got a real flair for the Old West, and to further enhance the Ponderosa-type atmosphere, he always has an old western flick playing on the tube. Once dinner was over, we unwound by watching TV, shooting some pool, and telling old war stories. Bruno has always been very kind to Jodi and me, but we've always kind of walked on pins and needles around him. He tells many a story about people he's angry with, and I have no desire to become a main character in his next tale. Besides, a good friend of mine long ago advised me that if we wanted to continue to have a good place to eat and sleep in town, we'd better never cross him. He's not only the sole proprietor of the restaurant, the bar, and the motel, but also several cattle ranches that spread across 75,000 acres. He's highly regarded throughout the area, and Bruno's is the only motel within a 70-mile radius to where we need to be. I've always taken that good advice to heart.

September 19, 2002, began with our prompt 4:00 a.m. arrival at Black Rock. There was a clear, dark-blue sky overhead, and I took a moment of silence to gaze upward to wish my mother well and let her know how much I missed her and that I hoped to make her proud this day. I felt it was truly the one I'd waited for—my dream and my vision were going to come true because our *Primera* rocket was going to make it into space, and we'd successfully open the door to help make space accessible not only to our fellow civilians but to companies of every size. Just as I was about to turn my gaze back to the tasks at hand, I spotted the moon shining on the International Space Center as it passed directly over our heads and felt it was a very good omen.

PRIMERA SPACE SHOT 2002

Left: Ky and Jodi on the launch tower.

Below: The *Primera* Space Shot 2002 team.

The Space Shot 2002 launch control center.

My jubilation was short-lived, however, as what had started out such a grandiose day soon became nothing short of an obstacle course. Our only possible launch time was from 9:00 - 9:20 a.m. At precisely 8:15, our RSO was notified that a train was scheduled to pass at 9:15, which was a direct infraction to the criteria allowed. The team flat out refused to give up and quickly decided to make a plea with Union Pacific. The phone call was made, and after a few minutes, they agreed to hold the train at the station for an additional 30 minutes so we could go forward with the launch and not jeopardize failing one of the conditions of our license. That was a first for amateur rocketry, and the countdown continued. No sooner had we overcome that obstacle than we were faced with another one. We needed to release three balloons at different altitudes in order to gather all the pertinent wind data required to properly position our launcher. The release of the 40,000-foot balloon took much longer than anticipated, making it impossible to get both the 10,000- and 2,000-footers in the air in the time remaining. We once again had to plead, and were able to reach consensus with the FAA, which agreed with our recommendation to delete the 10,000-foot balloon and rely on the readings of the third one to show convergence, which resulted in a 5-mile-per-hour error. A 7-mile-per-hour error or greater would result in a launch scrub. The countdown procedures were redlined in real time, resulting in a timeline change to begin countdown at 8:30 a.m., putting the launch at the tail end of our one-hour window.

At countdown, operation start propulsion was quickly a no-go, as the motor temperature was below the minimum operation limit of 55 degrees Fahrenheit. I gave the go-ahead, hoping the morning sun would help get the motor above the limit, and it did. At 9:10, the propulsion unit also gave the go-ahead, and the countdown continued. The firing solution elevation angle for the launcher was calculated to be 2.1 degrees off vertical. This was 0.9 degrees beyond the minimum tilt limit of 3. The launcher azimuth was 10.2 degrees, which was a mere 0.2 degrees from a no-go. The team scrambled to fill a backup balloon in hopes the winds had changed enough to bring the firing solution within bounds, only to find we were completely out of helium. I immediately wondered whether our friends from the Burning Man festival had decided to help themselves to our helium supply in an effort to get high. There was no other logical explanation, because the bottle was completely empty and had been full when we left. We quickly realized why so few of them had shown up to watch. The CSXT then recommended to the FAA that the launcher be set to the limit of 3 degrees, which would move the landing site 4 miles down range. The FAA privately discussed our recommendation and after five minutes gave the go-ahead. The BLM also cooperated and agreed to the change in the landing site. The countdown continued.

At T-minus 7 minutes, the wind speed drifted back in limits and the countdown proceeded with the final go, no-go poll. At T-minus 2, with only 6 minutes left in our window, the CSXT launch team gave the final go for launch. For safety reasons, everyone had to leave the launch control center, and this was it. We'd made it to final countdown, and in moments would be celebrating all of our hard work.

Launching a weather balloon to monitor wind conditions.

At exactly 9:15, T-minus 0, I grabbed Jerry's hand so we could push the button together, and as we did, nothing happened. The motor failed to ignite. We were absolutely stunned. How could this possibly happen when we'd come this far? The RSO was dispatched to the pad to check the firing line. His drive back nearly set a new land speed record, and at precisely 9:18, the T-minus 60-second terminal countdown resumed.

ROCKETMAN

At 9:19, with only 1 minute remaining in our window, Jerry and I once again pushed the button. Silence enveloped us as we stood with gaping jaws, watching our rocket with great anticipation as nothing happened. I was about to cry out in anguish when the *Primera* suddenly roared to life and silenced me. It cleared the tower, accelerating at 16 g's, and was performing flawlessly, flying straight and true, precisely on its planned trajectory and heading straight into space. The jubilation began once again. With all eyes to the sky and smiles of joy on our faces we watched and marveled, and just as we were sure we'd accomplished our mission, the smiles vanished, and the unthinkable happened. Our pride, our joy, and our dreams came to a horrific and abrupt end as we witnessed a huge explosion on board and watched the *Primera* shatter into thousands of pieces and then dissipate into a shower of particles falling quickly to the ground. We all ran for cover and stood helplessly by, experiencing the true agony of defeat.

Jodi and I and some members of the crew examine part of the wreckage of the *Primera* Space Shot rocket. At that point, my whole world started to fall apart.

PRIMERA SPACE SHOT 2002

How could this happen? The *Primera* had carried a big piece of my life with it, and I just started to cry. I'd been through this before, but this was exceptionally hard because we'd come so close this time. We'd truly given 110 percent effort, and it was all for naught. I knew the rest of the team was as devastated and disappointed as I was, and yet through all the heartache, helplessness, and pain, something truly wonderful happened to me that day. I believe with all my heart that my mother was there to catch me as I fell and to pick me right back up again. At the beginning of this attempt, I had vowed to call it quits if our mission failed, yet I somehow managed to walk away from the desert that day a very, very proud man. Somehow, instead of feeling totally beaten, my mindset was stronger than ever before, and I knew that the promise I'd made to my mother was one I would someday, somehow, keep. I would fulfill my dream one day, of that I was certain.

Everybody thought it was over, including me. But in my heart, I knew it was never over, even if I had to give up everything else in my life to accomplish my dream.

Chapter 25

The Journey Continues

I was emotionally drained and started having pains in my chest and stomach. I'd lie awake at night, worrying about the prospect of ever succeeding in my mission and what my future would hold. Was this truly the way I wanted to spend the rest of my God-given time here? On top of that, because I had spent the past several years focusing all my energy on fulfilling my lifelong dream, I had cut way back on earning an income and was spending more than I was bringing in. This brought on some financial difficulties, and I found myself struggling to maintain the lifestyle to which my family and I were so accustomed.

Ironically, I received a phone call one cold November day from my old friend, Fred Goeske. He called to relay a story to me about his *own* health, and it really made me sit up and listen. He told me he had been experiencing chest pains and that he had fortunately gotten to the doctor just in time. They'd found an 80 percent blockage in his heart, so critical that it required immediate surgery to prevent his early death. That conversation scared me into taking a close look at my own health. I no sooner hung up the phone than I had an appointment to see a specialist for a full physical the very next day. I sat on pins and needles for two full weeks waiting for the results of all the tests they'd taken, as I was certain something was wrong. The phone call finally came, and my doctor told me my cholesterol was a little high and that I had what was known as a Class A heart. "A Class A heart?" I pondered. I suddenly felt my ticker pounding very heavily in my chest, for that was the news I had so dreaded hearing. "I must be missing a heart valve or maybe even something worse," I thought to myself. With great apprehension, I asked, "How bad is it then?"

THE JOURNEY CONTINUES

He caught me completely off guard with, "Ky, a Class A heart is a good thing. You've got the heart of a 20-year-old man! There's very little deposit built up, and you can continue eating just about anything you choose for the rest of your life. There's a less than 10 percent chance of you, my friend, ever suffering a heart attack."

That was the best news I'd heard in a long, long time, and I jumped out of my chair as soon as I hung up the phone to go share it with my wife. With a huge smile on my face, I jokingly told her, "Jodi, I have but three goals in my life. The first is to be a good father and raise my two young children right, and hopefully live long enough to see them have children of their own. The second is for me to be able to fulfill my lifelong dream and successfully get my rocket into space. And the last one is for me to outlive *you*." My wife is a whole lot younger than I am, so we both got a good laugh out of that. Over the years, she and I have oftentimes kidded each other about the life insurance policy she has on me. This prompted me to ask her just exactly how much I'd be worth if I was somehow robbed of achieving my third goal. When she answered, I told her to go take one out on herself, too, for the same amount, and put it in our children's names. Don't get me wrong . . . I love my wife with all my heart, but the news of my good health just really kicked up my competitiveness, and I seriously felt like I'd already inherited the million bucks she was going to owe me.

I constantly thank the good Lord for giving me so many blessings in my life, but I felt the warm fuzzy feeling the recent clean bill of health had given me rapidly dissipating as the desire to fulfill my second goal once again overtook me. As the months passed, the stress of trying to get our space shot program back on track brought on some serious depression, which led to a significant overeating problem for me. The stomach pains I had overcome earlier were now more severe than ever, and I was once again unable to sleep.

That being said, and with my mind made up, we decided to take a vacation and get away from it all. I needed to spend some quality time with Jodi and just give myself some down time and shift my focus to other things. We decided to head down to Florida for Speed Week and see some of the other sites around the state. Just as we were heading down to Tampa, I received a call from Jerry. I hadn't exactly filled him in as to the decision I had made, so needless to say he was ecstatic and told me he had some really good news for me. Now mind you, Jodi was listening to the conversation, so I was trying to keep things on a real even keel and not let Jerry's excitement overtake me. He said a company named

Fuscient wanted to sponsor our next space shot! It was apparently a small company, but they were willing to give us enough money to get the machine work done and would also contribute toward supplies.

To say I wasn't absolutely delighted would be a major understatement, but I would definitely have to run this one past Jodi, and I wasn't exactly sure how she'd react. I thanked Jerry for calling and told him I'd have to give him a call back. Jodi asked me what was up, so I told her what Jerry had said and that it really shed a whole different light on things. She was actually somewhat supportive, and I found myself starting to get pretty excited about the whole thing again. But, as we continued to drive, I also started to worry about where we'd get the rest of the money to finish the rocket and the hundred other things we'd need to fly it. All of a sudden, that sickening feeling in my stomach overtook me again. I just couldn't believe I was actually considering letting myself get into another situation where the stress would be too overwhelming for me to handle. The whole thing really put a damper on the rest of our vacation, and I struggled with myself over making the final decision.

When we returned to Minnesota, I began having serious anxiety attacks and would toss and turn all night, completely unable to sleep. I felt like a big ball of nerves. Suddenly, life was not good. This went on for a couple of months before I finally realized I needed to do other things to help soften my emotions, or I'd likely wind up in the hospital. I decided it was time to try to have some fun, and I knew just the thing to make it happen. Of all the rocket toys I've built in my life, there was something I hadn't done yet that I was sure would bring a smile to many a face. I had heard about an upcoming national event that would be televised by a company called First TV. They were producing a show called the "Rocket Challenge." I knew it would be big, and I had the perfect plan for getting on the show. Crazy as it sounds, I wanted to build the first-ever rocket-powered outhouse (otherwise known as a Porta Potty)! I laughed to myself just thinking about it, so I called together a group of some of my wackiest friends to get their feedback. Not surprisingly, I instantly had an entire crew put together to make it all happen. Jodi was also very gung-ho about the whole thing, and her enthusiasm really helped to make it all the more fun for me. Satellite Industries is a local company in Minnesota that manufactures portable outhouses, so I arranged for a purchase. Needless to say, when I told the salesman what my plans were, he about split a gut. Hey, a sale's a sale.

THE JOURNEY CONTINUES

We researched the rules and regulations for the show, which would include several different types of contests. We opted to enter the one where the object was to see who could fly a rocket to the lowest altitude, deploy a parachute, and bring it back safely. I spent a couple weeks making modifications and assembled two rocket motors to generate enough power to lift it off the ground and make it fly.

As are all the toys I build, this was yet another work of art. My good buddy, Bruce Lee, came up from Omaha, Nebraska, to help me put on the final touches, such as the monkey we had strapped to the seat inside and the rolls of toilet paper we mounted everywhere! Our film crew flew in from California to document us getting ready for the big event, and they had as much fun laughing at this entry as we all did. Standing there admiring the totally *non*aerodynamically shaped bright neon orange and white oddity, we decided the only thing left to do was give it a name. In mere minutes, we dubbed it *Our Stinkin' Rocket*, and we were ready for liftoff. We got everything loaded into the trailer, and Jodi and I drove down to Kansas for the week-long event. I knew this national rocket launch would help get my mind off our space shot rocket, and it felt truly great to get away from home with my wife. The anticipation of not really knowing what would happen when

we fired the two rockets was also a good distraction, and I could hardly wait to see it. Because of its shape and size, I figured it'd be darn near equivalent to trying to fly a boxcar.

We met up with the rest of the team and proudly displayed *Our Stinkin' Rocket* for several days before our launch window. It was hilarious to hear the comments of everyone who came to see it. The two we heard most often were, "That thing will never fly," and "You guys are nuts." The more I heard, the more incentive I had to somehow make it work. When it came time to fire the rockets, I have to admit being a little apprehensive, and I know a lot of the spectators were looking for a place to take cover, just in case things didn't go right. Amazingly, though, when the countdown started and the firing button was pushed, *Our Stinkin' Rocket* flew off the launch pad in fine style. The dual parachutes came out a couple of seconds after it left the pad, with the 1,000-pound-thrust motors still burning. The flight was absolutely awesome! I know I never thought I'd see the day when I'd look up into the sky and see a flying outhouse! It resembled someone putting wings on a hippo, only not as graceful, and it was completely out of control. Without question, it was the most hilarious thing I've ever launched in my life, and like many others, it came to a crashing but survivable end. Much to our surprise, and to those who witnessed its flight, it was still in flyable shape for another day. The footage of this state-of-the-art, one-of-a-kind rocket was captured on film and broadcast on the Discovery Channel for thousands of people to enjoy, which delighted me to no end.

With very limited funds available, we slowly continued making progress on our new space shot rocket throughout the summer of 2003. As had happened with the good folks at Fuscient, we needed another sponsor to step in with some funding. We were pretty much at a

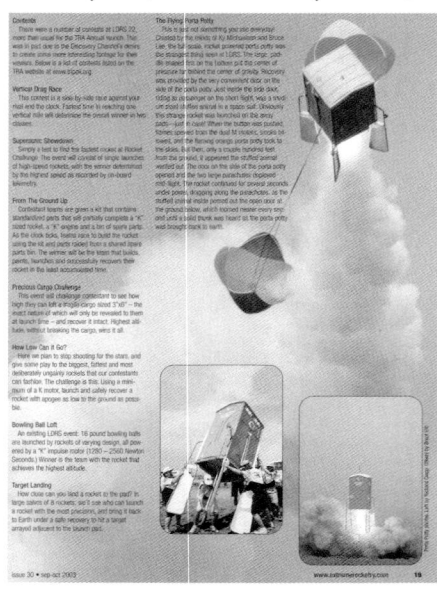

Our Stinkin' Rocket was featured in *Extreme Rocketry*.

standstill without it. As luck would have it, I got a call one day from my friend, John Hewitt, who happened to work for a company called Go Fast. John had called to ask for help with a promotion he was interested in doing for his company and wanted to know if I could possibly help him build a rocket belt. I knew the concept of such a device dated back to science fiction of the late 1920s comic strip hero, Buck Rogers, but it wasn't something I had ever really done before. I recalled old Buck used one to propel himself into the future, but I knew it wasn't all just a fantasy.

In real life some 40 years ago, an engineer by the name of Wendell Moore, who worked for Bell Aerospace, was credited with actually building the first successful rocket belt. He initially called it "small rocket lift device," or SRLD. That modern-day wonder was powered by a 90 percent hydrogen peroxide rocket that had a burn time of less than 30 seconds, so while it wasn't very practical for manned flight, it was definitely an attention-getter.

I was more than happy to help and work with him on the design, and I told John how to make a catalyst pack for the belt. We pretty much dealt with each other over the phone, so after a number of calls, I finally had to ask what Go Fast was. He told me it was a sports drink and that the owner of the company, Troy Wigsley, was a very promotionally minded businessman. It didn't take me one minute longer to realize that this guy could potentially become our knight in shining armor.

I asked John for Troy's number and gave him a call. I introduced myself and let him know I was helping John with the belt. Without hesitation, I continued sharing my dream and told him how hard we had all worked to achieve it, and how we were at a standstill now without enough funds to continue. He seemed very intrigued and wanted to know more. Well, it turns out that Jerry actually lived very near Troy, so I asked if it would be okay for Jerry to pay a visit to the company and bring some photos along of our past space shots. Troy very graciously agreed and called me a short time later with the best news ever. He informed me that his company would be more than happy to provide us with the additional funding we needed for our upcoming attempt. Those words were truly like music to my ears, and I couldn't thank him enough. In one of the happiest phone calls I have ever made, I got to inform Jerry that our Go Fast rocket would be built to completion.

With the rest of the crew informed of the new funding available, we were off and running again. Our plans were to fly in September or October of 2003, but just as we had completed updating most of our equipment and were close to finishing our rocket, another bombshell was dropped on us. The owner of the company who made the propellant, who was also a team member, sold his proprietary information and formulas to a self-proclaimed competitor of ours. This left us with a rocket but no propellant and absolutely zero resources for finding any. With the heaviest of hearts, we knew our space attempt for 2003 was over, and possibly over for good. I've no doubt they realized what a devastating blow this was to our team and the program, but they truly underestimated our will to succeed. That was something no one was going to take away from us. That's not to say people weren't going to try, because as it turned out, the propellant was only one hurdle; there was another one on a much larger scale for us to contend with.

We had some serious competition out on the West Coast when a company by the name of Scaled Composites, headed by Burt Rutan, decided to go after the Ansari X-Prize, which is the ultimate contest for rocket enthusiasts around the world. It's a race to get an astronaut and privately owned rocket, able to carry three people, into space (defined as 100 kilometers above Earth), return safely, and repeat the feat within two weeks. The prize for doing so, which cannot be claimed unless the team gives notice of the attempt and carries three passengers, is $10 million. Well, needless to say, we all knew that event was on a much different level than what we were trying to accomplish and felt it would, sadly, take a lot away from both our accomplishments and theirs.

Everyone remembers *Sputnik*, the world's first artificial satellite, but nobody remembers the second one to achieve the same thing, and I wasn't going to give up my dream because of some wealthy competitors.

Their project was solely financed by Microsoft's cofounder, Paul G. Allen, who is one of the richest men in America. We had all been keeping close tabs on their progress and knew they'd built both the spacecraft, named *SpaceShipOne*, and its aircraft carrier, *White Knight*. The carrier was a beautiful all-composite twin turbojet designed to lift the spacecraft from the runway to roughly 50,000 feet before releasing *SpaceShipOne* into a glide, at which time its pilot would fire his rocket motor and reach mach 3 in a vertical climb. Scaled Composites is a hugely successful aerospace and specialty composites company, responsible for the Voyager, the first airplane to fly around the world without refueling. When I said we had serious competition, I wasn't kidding!

In retrospect, however, one of the most awe-inspiring revelations about our time here on this Earth is that history doesn't care about any of those things. History isn't defined by *how* something has been accomplished so much as that it *was* accomplished. Everyone remembers *Sputnik*, the world's first artificial satellite, but nobody remembers the second one to achieve the same thing, and I wasn't going to give up my dream because of some wealthy competitors. When I first started my quest for space in 1997, there were about 25 other teams in this country who had the same dreams as I did but became discouraged and gave up hope. I was determined *not* to let that happen. I knew how many talented people had been on those teams, so I began researching previous members in hopes of finding someone who could help us out.

As I went down the list, my thoughts turned to Korey Kline, a fellow rocket enthusiast I met back in 1984. I knew he was still very active in high-powered rocketry and had in fact developed and patented a rocket motor, called a hybrid, that ran on nitrous oxide and plastic. Since we had no way of getting propellant for our previous motor, I realized a hybrid was the answer, but it meant starting over almost from scratch and building a whole new rocket. That wasn't exactly the best scenario, since we had spent almost all our money on the last one. However, when you firmly believe in something with every ounce of your being, the way I believed in this project, things like that truly become menial, so I just gave it some serious thought. I concluded that all we needed to do was slow down, think, and take the rest of the project one step at a time. Step one was finding someone to build our new motor, and step two would be the funding, in that order. The thought of starting over was overwhelming, but there was no way I was giving up the ship, not now, not *ever*. If someone else beat me to space, I'd be breaking my promise to my mother, and my dream would be just that . . . only a dream.

I called Korey and told him the situation we were in. I went on to reiterate how I knew we both had the same goals, and that maybe instead of working against each other, we could team up and work together.

He replied, "Do you mean to say you can't make it without me?"

"No, Korey. You can't make it without me, and I can't make it without you." I continued, "Now do you get the real picture?"

With just a moment's hesitation, he replied, "You're right. Let's team up!"

We talked a bit longer and came up with another team member and fellow rocket buddy, Derek Deville. We added Derek to the conference call, then brought Jerry on, and unanimously decided we'd all just have to continue funding the project on our own. It wasn't long into the conversation that I knew bringing Korey and Derek onto the team was a real good decision, and we agreed the hybrid rocket technology was the way to go. I also figured this was it—our last chance to become the first civilians in space. I knew the competition was lurking out there, but I for one wasn't going down without a fight.

Over the next two months, the team went all out to design a motor to fit our needs. The hybrid technology meant the rocket size had to increase, and it just kept getting bigger and bigger, to the point that when we ran the numbers on the computer, it started appearing we might not even be able to launch at Black Rock. It just didn't have the acceleration and was too wind-sensitive to get off the pad quickly enough to become stable. We knew from past experiences how wind could be our worst enemy, so this thing was turning into another catch-22, and I became extremely frazzled once again.

It was all getting to me. I didn't think we had to reinvent the wheel to make this rocket work, but it was definitely starting to look that way. I called Jerry and told him we needed to stop beating our heads against the wall, and go back to our ammonium perchlorate (AP) motor. The only way we could do that was to bring another fellow enthusiast on board, Chuck Rogers, who specialized in numbers—specifically, rocketry physics. So with Jerry's blessing, we gave him a call and filled him in on the situation. Once an enthusiast, always an enthusiast. He was in and hopped right on board.

We decided to do all the design work on the computer first, and with Chuck's many years of experience at building hobby rockets, it turned out to be a great idea. Then we used Derek's formula, which he called D8, to make modifications to our design. From there, we started building small rocket motors that we put on test stands and fired. The information was then fed back to Derek, Jerry, and Chuck. Our hopes lifted as the initial test fires were all very positive. The next phase was to scale up from a 3-inch to a 6-inch diameter, which resulted in a motor with characteristics that really looked promising. From there, we scaled up again to a length of 175 inches, with a 10-inch diameter—a monster by anyone's stretch of the imagination and absolutely without question the largest amateur rocket motor ever built. The solid-propellant motor contained 435 pounds of AP-based propellant configured in a monolithic case bonded grain with a central fin-o-cyl core and a nearly neutral thrust profile. The case was made out of 6061 aluminum, with the end closures threaded to the casing. The nozzle was created from a new process using a combination of graphite, carbon-fiber, and ablative materials and featured a bell-shaped exit cone. The motor delivered just over 104,000-pound-seconds of thrust, with an average thrust of 12,000 pounds per second.

In the past, we'd relied on a trusted member of our team to build an engine that wouldn't fail; yet, I always felt he had used us as his guinea pigs to test out his designs. The previous motor had in fact failed in flight, and when we needed him to once again build a new engine for us, he had sold his technology and formulations to a competitor of ours. In hindsight, I realize now that what I had thought was such a travesty at the time had truly turned out to be a blessing in disguise, because I now had Korey, Derek, Jerry, and Chuck on our team. We were all convinced that what had happened in the past wasn't going to happen again, because this time we were going to test our rocket on a stand before we used it in flight.

It was now March 2004, and plans for the test were set in place. We decided on a secluded location in Florida, so Jodi and I flew out to meet the rest of the team there. However, when we arrived at the airport, and I turned on my cell phone, it rang almost immediately. Jerry was calling from Colorado telling us *not* to fly out, because Derek's wife was in the middle of giving birth to twins. I told Jerry, "Too late, man! We're standing in the middle of the Miami airport!" We both got a big laugh out of that one. I knew this meant a push back for the test, and that now Jodi and I wouldn't be able to attend, but we decided to make the best of the situation anyway. After all, we were already there, so we checked into our hotel and then headed out to Alligator Alley, where we rented an airboat and spent our day chasing alligators. Interestingly enough, we had a chance to feed them what we learned was their favorite snack—marshmallows. You truly learn something new every day.

> **I knew the competition was lurking out there, but I for one wasn't going down without a fight.**

We returned home in just a couple of days so we could spend the Easter holiday with Jodi's parents, who were coming down from North Dakota. Everything went great, and we enjoyed their visit immensely—that is until my phone rang just as they were getting ready to leave. It was Monday, April 12, and it was Jerry calling from Florida. The team was about to perform the test, and they wanted me to at least have the chance to hear it. I hopped up and caught Jodi and her parents as they were heading out the door and very excitedly told them what was about to happen, and that they should come back into my office and listen over the speaker phone. I truly can't explain how excited I was, yet from somewhere deep down inside, I had a serious inkling that something was going to go very wrong. Before I could say anything, I heard Jerry give the countdown. Seconds later, we heard this huge roar that lasted for less than one second and then quit.

I knew from past experiences that something had just gone horribly wrong, and that something terrible had happened, but I couldn't figure it out from 2,000 miles away. Tears started rolling down my cheeks as I heard Jerry say, "It's under water. It's under water, and there's water and fire shooting 30 feet in the air!" I could hear people scrambling, and I just sat in silence and hung up the phone. This was an unbearable letdown, and my mind started racing. I became extremely emotional and couldn't help but question whether or not this was truly the end of my dream.

All those thoughts were racing through my mind, but I was soon snapped back to reality when Jodi's mother started in on me. She said some incredibly harsh things that I really didn't want or need to hear, but she persevered with, "When are you going to stop spending all that money and start thinking about your family's future?" That statement cut me deep to the core because anyone who truly knows me knows how much I love and cherish my family, and that there's nothing I wouldn't do for them. I took her words to heart, but also maintained to myself that what I was trying to do wasn't just for me; it was also for my family's security and to fulfill the promise I made to my mother on her deathbed. I had to let it bounce off. I just had to, and the setback I had heard over the phone only made me more determined to end the nightmare once and for all and finish the job at hand.

Our farewells obviously weren't the best, and Jodi was outwardly upset with not only the whole situation, but especially me, yet I just took it all in stride. I had Jerry e-mail a copy of the video to me so I could see it for myself and try to figure out what had gone so wrong. The motor was bolted down to a large steel I-beam, and the front was up against a load cell to measure the thrust. When it was fired, the threaded end closures on the casing failed, blowing the forward closure off. Instead of the motor pushing up against the load cell, the thrust went in a different direction, which in return tore the motor off the bolts holding it to the stand. The rocket flew right into a rock quarry filled with deep water. The video truly painted the picture, but we wouldn't be satisfied without a recovery, so Korey brought in some of his scuba diving buddies a few days later to use lifting bags to try to retrieve it. They were successful in getting it out of the drink and onto shore, and that's when they clearly saw where the failure was. The power of the motor had overwhelmed the threaded closures. It was obvious we needed to make a significant design change. We took a couple of days off before Jerry, Derek, Korey, and I had another conference call.

"Is this the end?" I asked first.

After a slight pause, they all piped up and replied, "No."

"We've come too far, guys, to stop now," I replied.

Derek said, "We need to make two more motors, one to static test, and another one to fly. We've got less than a month before our first scheduled launch day."

In stony silence, I secretly told myself this was darn near impossible. The fact of the matter was, though, that the entire project had not been easy on anyone, and we'd been snake bitten from the start. Unfortunately, the pattern continued, and the bites got harder each time. We ended our conversation in total agreement that we were going for it one last time, and each of us had a list of things to do. Derek and Korey ordered the chemicals, and Jerry ordered the aluminum. From Minnesota, I started back at square one in hopes of finding someone to help us build the two motors. I was lucky enough to find a machinist who needed some extra cash and was willing to work for us at night when he got off his regular job. Out on the East Coast, fellow team member Eric and his crew were finishing up the avionics and getting them ready to ship out to Jerry for testing. To make things even more hectic, several other groups with multi-million-dollar budgets and the same ambitions we had were nipping at our heels to get into space first. The ultimate low blow came, though, when we learned that one of those teams heard we had been talking to the BLM about launching our rocket on May 24th, so they blocked out that date for themselves to ensure we didn't get it.

Regardless of whether or not they truly intended to make a launch attempt that day, they made their move to prevent us from going first. That prompted me to call Jerry immediately. We had a long talk about the sudden pressures put upon us, and I told him we simply had no choice but to bypass testing another motor. Time was no longer on our side, and we just needed to fly. Jerry agreed, so it was decided unanimously that the only motor going with us out to Black Rock would be our test motor. We also decided to back up our launch date to May 17, which meant we would have to roll the dice and take our chances.

With less than three weeks to get everything done and be in our launch tower by the 17th, we realized the odds of us making it were less than 10 percent. I've never been a gambling man, yet those odds sounded nearly impossible to me; nonetheless, we were a go. This decision required major team effort, with everyone giving 120 percent, and Korey and Derek really came through like superstars. They worked late night after night mixing and pouring the propellant into the rocket motor, and everyone shifted into high gear. We put together the plan of all plans that we knew would catch everyone off guard. We told no one, and I mean *no one*, but the team members when the launch would be.

I bought a new trailer to haul the motor on, and my friend, Matt Murphy, and I drove straight to Miami from Minneapolis. The first night on the road we made it to Illinois, but the battery shorted out, so we ended up sitting on the roadside for over two hours waiting for help. There had apparently been a serious accident down the road, and all the emergency people were tied up. When help finally arrived, it was late, and we were tired. We had the tow truck get us to the nearest small town, got a motel, and headed to the local K-mart first thing in the morning for a new battery. We started down the road once again, but our journey was disrupted about two hours later when our alternator went out. We limped along to the next town, but the parts store was closed until 10:00, so we had no choice but to wait. I remember sitting in the parking lot, praying they'd have an alternator to fit the truck. We got inside the store, and my prayers were answered. An hour later, we were back in business. Two days on the road, and two days' worth of trouble. I tried really hard not to think about whether or not this was a sign of more bad luck to come, but it was definitely in the back of my mind. In spite of our troubles and added pressures, we made it to Florida in record time and headed straight to the place where the motor was built so we could assist in getting it ready to transport. After careful preparation, we loaded it up and took off across country en route to Nevada. We needed to drive 6,000 miles in six short days, so we crossed our fingers, hoping the truck would get us through.

That night, exhaustion began to set in. We were on a tollway but didn't see any signs for hotels or lodging of any kind anywhere along the way. We decided to chance it and take the exit marked Heehaw. It wound around to

THE JOURNEY CONTINUES

another tollbooth, where I asked the attendant if she knew of any hotels or motels in the small town, and she replied, "Yes, there's one . . . but you don't want to stay there." I thought that was kind of a strange answer, so I asked her why she would say that, and she continued, "Ever hear of the Bates Motel?"

"Why, was someone murdered there?" I asked.

She just laughed. We drove into Heehaw, and stopped to get some gas and a sandwich. As the young lady was making our sandwiches, I asked her if there was a motel in town.

"You see that yellow light about a block down?" she said, pointing down the road. I nodded my head, and she continued, "Yeah? Well I wouldn't stay there if I was you."

I said, "What's going on, anyway?"

"Look, I'm just a little girl from a town called Heehaw." With that, she handed us our sandwiches and walked away.

I was so tired, I just about died laughing. It wasn't too long before the girl at the cash register, who'd been listening to our conversation, came over and told us she wouldn't stay there either.

We'd heard enough, so tired or not, we left the wonderful little town of Heehaw and its deep dark secrets behind us, and ended up driving another two hours before we found a safe motel to sleep at. It was now May 11, and Matt started out driving. He decided we'd better fuel up, so just as he was pulling into a gas station, we heard a big *bang*. He'd hit a steel pole protecting the pumps with the fender on the trailer. I guess things don't stay new forever. Everything went smoothly from there, until the little screw from my eyeglasses fell out and took my lens along with it. I started panicking, because I can't see a thing without them, so he pulled over along the edge of the freeway to try to find the little screw. Needless to say, I was no help at all, but thankfully he found it and we got over that little dilemma.

> **I knew from past experience that something had just gone horribly wrong, and that something terrible had happened, but I couldn't figure it out from 2,000 miles away.**

My cell phone started ringing off the hook, so I got busy handling all the calls and lost track of Matt's driving for quite a while. After the last phone call, I asked him if he knew where we were, and he said, "No." I picked up the map and informed him that he'd driven us 100 miles out of our way. He looked at me kinda funny and said, "Actually, we're 200 miles out of our way because we're 100 miles south of where we're supposed to be."

I had really planned on stopping early that night for a nice dinner so I could satisfy my overwhelming craving for some Belgian waffles, a 2-inch steak, and a good night's sleep. So much for that—we ended up driving another four hours. As we were cruising along, something dawned on me. I asked, "Hey, what'd you say the date was today?"

"It's the 11th."

"Wow," I marveled, "The 11th has been my unlucky number for over 40 years." I spent roughly an hour reminiscing about all the bad things that had happened to me on the 11th, which somehow helped the time go by, but it was nearly 2:00 a.m. before we finally stopped for the night. We racked up 1,200 miles, 100 miles south, 100 miles north, and 1,000 miles west. It struck us as really funny because Jerry was sending e-mails to the team encouraging them to watch us as we traveled to our destiny. Good luck with that one.

The fifth day into our trip, the most bizarre yet wonderful thing happened to me. My phone rang, and a voice on the other end identified himself as Don Gray of Lake Havasu, Arizona. He'd apparently called my home and had gotten my number from Jodi. Anyway, I had no idea who he was but was instantly intrigued when he told me he had some information on the Michaelson motorcycle—a vehicle my family had built in Minneapolis early in the twentieth century. I was very interested in it but little information survived. With undivided attention, I listened intently as he told his tale. He'd recently found an old trunk that had belonged to his grandfather, and in the trunk he'd found lots of old pictures. There was a picture of what he believed was a 1912 Cadillac and another one of his grandfather on a motorcycle. He could plainly see the license plate number on the car, so took the photos to the historical society. He then set about sifting through a pile of documents and found an envelope containing the original registration for the Caddy and the registration for a 1914 Michaelson motorcycle. He studied the picture under a magnifying glass and was able to make out the Michaelson logo on the gas tank. That prompted him to do some research on the Internet, and that's how he found out my family built the motorcycle. I was beyond words and just couldn't

THE JOURNEY CONTINUES

believe this man had taken so much time and effort to share this amazing discovery, having no idea what it truly meant to me. What a small world indeed.

He asked me where in Minnesota I lived, so I told him but let him know I was on the road and would be going through Arizona the next day. We discovered I'd actually be passing about 100 miles north of him, so he asked me if I'd be able to stop in for a visit. He said he'd give me copies of the photos with the motorcycle. I told him I'd do just about anything to have them, but I was on a mission, and I just couldn't stop. He went out even further on a limb for me and told me he'd be happy to meet me halfway at a truck stop on I-40 in Arizona. I couldn't believe it! What a truly wonderful person he was. We made arrangements, and the anticipation of meeting Don was almost as taxing as making it to our launch site. I met him there at the truck stop and had the honor of telling him to his face what a great guy I thought he was for sharing his family's history with my family's history, and what a coincidence I found it all to be. It was truly a one-in-a-million long shot I'd have ever gotten my hands on those pictures. One of the photos showed his grandfather next to the Michaelson motorcycle on the Canyon Padre Bridge that was being built for Route 66, and the other ones were by an old log cabin. In our short time together, I learned that his grandfather was one of the original pioneers of Flagstaff, Arizona. I could have stayed and talked to him for hours, but the clock was ticking, and we simply had to move on. We said our goodbyes, but our paths crossed for an eternity that day.

Above: The 1914 Michaelson motorcycle.

Right: Don Gray and Ky Michaelson, Arizona, 2004.

Chapter 26

Three Days to Countdown

Jerry put the word out to the team to pack up and leave for Nevada, because Matt and I would be there Saturday. Our friends from Fuscient and Go Fast, as well as many other sponsors we'd had over the years, were also en route to witness the big event and to lend us moral support. The plan was for all of us to meet out in Black Rock on Sunday to assemble the rocket, leaving Monday open for our first launch attempt.

Matt and I checked into Bruno's on Saturday in really good spirits but physically exhausted. Bedtime came early that night, but excitement kept me from getting a real good night's sleep. We met for breakfast at the break of dawn, then headed straight to the desert. The anticipation of meeting together as a whole group and a team for the first time was incredible, and I just couldn't get us there fast enough. I cannot begin to tell you how proud I was of everyone who showed up to share a dream with me, and how tickled pink I was knowing that parts for our rocket had actually come from six different states. I was bursting with confidence that this was going to be our day, and for once things were going to go our way. We deserved it, we'd worked hard for it, and we'd paid our dues and then some—this was our time to shine.

Space Shot 2004, Black Rock Desert.

THREE DAYS TO COUNTDOWN

Putting decals on the Space Shot 2004 rocket.

We finished assembling the motor, which the team covered with a full Go Fast decal before attaching the red fin can. Fuscient and several of our other past and present sponsors added a "USA" and American flag logo right beneath the red nose cone. Those decals and the painted-on white, silver, and red Go Fast! and Fuscient logos gave its sleek black body a stunning look. It was a bona fide work of art and a thing of beauty. If only my mother and dad could see their son and know how proud he was. I felt like nothing could stop us now. With everything coming together so quickly and going our way, we all decided this seemed to be the day we'd waited for, and we didn't want to wait until Monday.

Everything was going smooth until John, one of the crew members, came over to me and said, "Ky, we've got a serious problem. They sent us the wrong detonators."

The detonators are used as a part of our recovery system. This was our first setback and was an enormously significant piece of our flight sequence. We were going to use an explosive charge to cut the rocket in half, which would slow it down enough to deploy a recovery parachute. That was critical because we absolutely had to recover our payload section, as it carried all the avionics and flight recorders. The technical data received via radio transmissions was our only other source of truly proving how high and how fast the rocket had gone. Without the parachute safely returning the payload section to us, we'd be left to rely solely on computer simulation, which would be far from accurate. We had no choice but to go back to Gerlach to try to get some help. Bruce Lee, Bruce Kelly, Jodi, and I agreed to make the trip. I knew what we needed, and who may have it, so I headed to the gas station owned by Bruno's son-in-law, Cecil. I very excitedly filled him in and told him I needed a 9/16th tap and a drill. We hunted and hunted between waiting on customers, and it took quite a while for us to find the tap, but he didn't have a drill. The nearest town was 120 miles away. My knees knocked as I explained with desperation that a round trip like that meant we'd be unable to put the rocket in the tower until Monday, and there'd be a good chance we'd lose a launch day because of the afternoon winds. We wanted to launch for sure by Monday morning if we could. He was very sympathetic, and we just stood there together, trying to come up with an alternative.

As we were talking, there was a guy filling up his Harley with gas. He walked over to us and said, "I didn't mean to listen in on your conversation, but I think I can help you guys."

Mind you that when I get excited, I talk very loudly, so I kinda laughed to myself at his remark, since I'm sure everyone within a mile from the gas station could hear me.

Nonetheless, he went on to say, "Give me 30 minutes, and I'll be back with the right-size drill bit so you can finish the job."

I shook his hand and assured him I'd be more than happy to wait. Sure enough, he came back with the right bit, and in 15 short minutes, the job was done. As we headed back to the desert, I radioed Jerry to contact the FAA, 'cuz we were now a go. When we got back to the launch site, we were less than one hour from sunset and wanted to take pictures of the rocket in the tower. While we were gone, Jerry had loaded up the payload section with American flags, post cards, and lots of other goodies that hopefully would be flown to space. I packed the parachutes, slipped on the payload section, and Jerry bolted it into place. The rocket was ready to fly. We spent the next 10 minutes taking pictures, then pushed the rocket on its carrier out to the launch tower and ever so gently slid it into place.

Billy and Ky, with the help of the crew, move the rocket toward the launch tower.

Now the real work began. We had to lift both the rocket and the tower into position, a combined weight of over 2,000 pounds. With a lot of grunting and groaning, soon the Go Fast rocket was standing proud and tall. Our surveillance helicopter landed 100 feet from the launch pad, and in all my excitement, I started screaming, "Look at that rocket. Man, that's a rocket!" Seeing it standing tall, and knowing all the hard work, sweat, and tears that had gone into it unquestionably made that the proudest moment of my life. Two FAA representatives, Michael Aherne and Marshall Phillips, arrived on the scene to make sure we followed all the rules and procedures by the book. For the first time since our involvement with the FAA Office of Commercial Space Transportation, I actually felt at ease with them. These two were extremely upbeat, which just added further excitement to the air. It had been a mutually long journey for both sides, but I sensed their feeling of ease with us as well.

What really put the icing on the cake was when they came over and said, "Nice rocket. This time you guys are going to do it."

Team member John Gormley on the launch tower.

To me, that was really heartwarming and helped set the mood. With everyone in position, we thought the time was perfect. But, as had happened so often in the past, Mother Nature thought otherwise. We half-heartedly stood immobile and in silent anguish watched the fiery orange glow of the sun set quickly over the mountaintops, leaving us no time to perform the mandatory dress rehearsal prior to launch. As it made its hasty exit, darkness, pouring rain, and lightning came in with a vengeance, and the whole atmosphere was transformed in a matter of minutes. We went from hope to worry, as we prayed that lightning wouldn't strike our launch tower and destroy all our electronics. We knew there would be no launch that day. A bit disheartened but still optimistic, we packed up our gear and headed back to town.

Chapter 27

Rend the Sky

There was no sleeping that night—only lying awake watching the minutes tick by on the clock, as if time had suddenly stood still. We were all out on the lakebed before the morning's sunrise, donning our bright red Go Fast/Fuscient–logo short-sleeve shirts and lookin' good. The sun made its warm welcome as it slowly crept higher and higher over the mountaintops, illuminating the once-dark sky with its majestic orange and yellow hues. The air was crisp, the sky was clear, and the winds were calm. I was filled with serenity at the beauty surrounding the area, and I knew in my heart this was the perfect day. Nothing was going to take away our moment of glory—*nothing*. We went through our government-required dress rehearsal in about an hour, and it was flawless. Everyone was in the right place at the right time, and we were operating like a well-oiled machine. We immediately positioned ourselves for launch mode and, speaking into the microphone, Jerry announced, "Good job. Now we are going into operational launch mode."

Billy Robin

REND THE SKY

As I watched Bruce Lee and Jodi start filling the weather balloon, I had a sudden attack of the butterflies, and the adrenaline kicked into high gear. As the old saying goes, I started shaking like a Harley rider. I radioed Salt Lake, Reno, and Sacramento air traffic control to open up our launch window from 11:00 to 11:30 a.m. Just as I completed that task, we suddenly received a call from our train spotters, who told us there was a maintenance truck on the track about 10 miles away.

I turned to Marshall Phillips and asked, "Is that considered a train?"

Thankfully, the reply was, "Nope."

I started grinning from ear to ear, knowing if there was a maintenance truck on the track, there was a pretty good chance no trains would be coming along anytime soon. I looked up to the heavens and saw a sky bluer than the ocean, with just a few scattered fluffy white clouds floating freely overhead. My spirit soared, and I was overcome with inexplicable joy. As soon as we received the wind data, Jerry ran the numbers on the laptop computer. John, the launch tower crew, and I jumped into our trucks and drove out to the rocket pad to wait for Jerry to give the final settings for the tower alignment. That operation is very vital, as one degree off could make the rocket land outside of our impact perimeter, which would violate our launch permit.

The tower itself is extremely heavy, so we used two big hydraulic floor jacks to maneuver it. Per Jerry's instruction, and with the help of four crew members, we tilted it to 7 degrees off vertical, which is nearly our limit. The maximum allowed is 8. That piece took about five precious minutes. We began experiencing 10- to 15-mile-per-hour wind gusts at the tower, so I was concerned the winds could impact us as they had in the past. As I nervously watched the skies, we heard Jerry say, "Everything is within limits, and we are green. Clear the pad!" The tower crew left, and it was just Jim, John, and me. John armed the rocket and hooked up the firing wires. As he did so, my heart started pounding profusely, and my whole body began shaking uncontrollably. My fingers were also tingling, and I thought to myself, "Calm down, man. This is no time to have a heart attack."

John looked at me and said, "Are you all right?" I could barely speak, so he grabbed my arm and told me, "Get in the car. Let's get out of here."

As I jumped in, he looked at me again and with a big smile said, "Ky, relax. I have something for you. I've been saving this for a long time, just for you."

Jim Hoffman, Ky Michaelson, and Jerry Larson in the Space Shot 2004 control center.

I nervously looked over, and he popped a tape into the cassette player. He pushed the play button, and it was a song from Metallica called "Nothing Else Matters."

He jokingly said, "Now, just sit back, relax, and most importantly, listen. We have plenty of time to get back to the flight line."

Surprisingly, the words to the song carried me away and really mirrored my feelings. "What I had to go through . . . what I had to give up to get here . . . never cared for what they do . . . never cared for the games they played . . . life is ours, we live it our way, like forever trusting who we are, and nothing else matters." The song spoke truths to what I had been thinking and took me to a very good and positive place.

As the car stopped and we got out, my emotions were running higher than ever as I ran over to the launch controller. Just then I heard someone holler out, "Someone or something is out on the playa!"

REND THE SKY

Liftoff!

We could see something in a mirage that turned and was headed toward us. Everyone strained to see what it was, and in a few minutes we were able to make out two motorcyclists speeding right at us. I prayed they'd keep coming and not head off in some other direction. We all started frantically waving ours arms and hands for them to come to us, which they finally did. The few minutes it took them to get there seemed like an eternity, and we were able to see it was a man and a woman, racing to see whose bike was fastest. They pulled up to see what was going on, and we hollered for them to stay put. Talk about another heart-pumping, launch-stopping experience. Thank God it was only temporary.

Just then, Jerry's voice came over the PA system saying, "RSO, one minute train and sky check." He then announced, "Everyone stop all communications except for emergency hold, if necessary." What he said next, although not meant to be, was terrifying to me. "T-minus 30 seconds check. Note the time is 11:10 a.m." Those final words before continuing countdown hung in my ears until they burned. As he left the control center to stand outside in case of falling debris should the rocket fail, my mind kept repeating that if we launched in 30 seconds, it would be 11:11 a.m. As I mentioned before, I've had nothing but bad things involving the number 11, and now it was a double digit. I about panicked thinking this was bound to become yet another string of bad luck and instantly went from optimist to pessimist. I knew I couldn't stand another failure, not at the age of 65.

From outside, Jerry gave the last commands before it was time to press the launch button. "Enable main switch," and then, "Arm vehicle." Once again, I began shaking uncontrollably, and tears welled up in my eyes. With all my might, I blinked them back because I needed to focus with everything I had. At that very moment, as I tried so hard to concentrate, all I could think of was how this had taken over seven years of my life and what I'd given for this moment. I knew no matter what happened in the next few seconds, it was the end. I was not going to put my family or myself through any more of this madness.

Jerry's countdown was to T-minus 10 seconds, resume count, when I reached for his hand. I wanted his hand on the button with mine because of all the hard work he had done; he was with me from the start. I was looking straight ahead at the rocket standing there so mightily in the launch tower, just gleaming, and could see the yellow light flashing, indicating it was armed. All I had to do now was push the button, but Jerry wasn't beside me. I didn't know his microphone wire had gotten caught on one of the cameraman's camcorders, and he couldn't get to me when the count was T-minus 0 . . . *fire*!

REND THE SKY

I waited about two seconds and pressed the firing switch. Nothing happened for a moment, but suddenly there was an enormous roar. Flames spewed from the huge 15-foot motor, and the rocket was pulling 23 g's as it lifted from the tower, leaving the desert floor behind in a gigantic whirlwind of sand and dust. The fiery flames, over 45 feet long, emitted an ominous puffy white jet stream that carried the rocket straight upward at over 600 miles an hour in the first two seconds!! The sound of liftoff echoed throughout the canyon walls and brought the whole desert to life! Tears of utter joy flowed like waterfalls as they gushed down my face. I stared straight up and started screaming out the burn time of the motor, "1, 2, 3. . . " It burned full power for 10 long seconds and, as it entered into the deep blue outskirts of space, was traveling close to 4,000 miles an hour: mach 5.5. As I began once again screaming, "Space, Space, Space," I knew for certain a big part of me was headed there, right along with it. The ultimate feeling of jubilation hit me, and once again unable to contain myself I shouted, "Jerry, you gave me the dream. You gave me my dream, man."

As I watched it disappear from sight, I yelled out, "Mom, I told you I could do it. I told you I could do it!" Every emotion inside me just poured out, and I felt as if I were being reborn. Even Jerry, who's normally a really calm, collected guy, and his young son, Eric, were jumping about 3 feet off the ground, and the whole crowd went absolutely wild.

We did this in front of our peers and friends, and there were cameras everywhere capturing the moment for the whole world to see. Jodi rushed over to me with tears flowing down her face, put her arms around me, and said, "I love you, Ky." Even through all the financial struggles and the stress and the strain, she stood by me through some pretty rough times, and she relished this moment almost as much as I did. Her words melted my heart on the spot, and I held her as close to me as I could.

The rocket was in flight and was in microgravity for about three minutes. Around seven minutes into the flight, John said he received a signal that the detonators had gone off, and the rocket was cut in half with the pyro charge. However, he wasn't sure if the parachutes were out properly. Because of the high speed at which they were deployed, there was a chance they could have torn right off the rocket's booster or payload section. Jerry announced, "Everyone please be quiet. If the parachutes didn't open properly, there's a good chance we'll hear two sonic booms." No sooner had he said that then we heard, "boom, boom" off in the distance.

We all concurred that the rocket had most likely hit the ground at supersonic speeds and had plowed so deep into the Earth that we'd never successfully recover the precious payload section, meaning we had no way of proving how high it had flown. As we stood around in a somewhat disappointed frame of mind, we suddenly heard another sonic boom, which made absolutely no sense to us at all. We thought the first two had come from the payload section and then from the booster, so where did this one come from?

Without time to further ponder the situation, one of the trackers who'd been standing there with his tracking antenna aimed straight up in the sky excitedly announced, "I've located it. I've located the payload. It's falling too slow," he continued, "It has to be under the parachute."

I couldn't believe it. Not only had we just made space history, but there was also still a real good chance we could make a full recovery of the payload section, and that our vital instrumentation and space memorabilia might still be intact. The reconnaissance helicopter took off to see if they could locate it as it was coming down. The tracking antenna was pointed toward the northwest, which led directly to a huge mountain range. I don't personally know any mountain climbers, and there were definitely none amongst us, so I knew this recovery effort was not going to be an easy task. As he pointed the antenna toward the top of the mountain, the tracker said, "It's not up there; it's on the ground."

Another tracker, Ian, of the Strato Fox tracking crew, then said something very inspirational that sent shivers down my spine. "We just made Black Rock the Kitty Hawk of the twenty-first century!"

Celebrating the success of Space Shot 2004.

That being said, I knew if I'd have flapped my arms, they'd have flown me like a bird out to the launch tower, but we opted to drive there instead. We had to see what was left of it after the inferno we'd witnessed during liftoff. I couldn't believe all the damage! It was bent like a pretzel. The lower part was so scorched a couple of the guys remarked that it looked like scrap iron and needed to go to the dump. I said, "Absolutely no way. That's a piece of space history, and it's coming back home with me."

Just then, the winds started picking up, and in a matter of just a few seconds, we found ourselves in a full-blown sandstorm that, as in the past, would have prevented us from launching just a few hours earlier. It didn't stop a few of the recovery guys, though. They were on a mission and decided to go ahead and start the search. The rest of the team and I went back to Bruno's to celebrate, which we did into the wee hours of the morning. However, I wasn't totally satisfied yet and knew my personal journey to space was not complete without that payload section. I knew there would always be skeptics out there and wanted assurance that we could undeniably prove to them and the rest of the world what we'd accomplished. That data rested somewhere up in those mountains, and that thought kept me wide awake yet another night.

Jim Hoffman stands to Ky's right, as Ky and Jerry Larson check out the flight data.

About 30 people showed up at the playa bright and early the next morning, and the guys who'd started the search the day before had gotten a pretty strong reading from the beacon on the rocket that did in fact seem to come from a mountaintop about 20 miles from the launch site. We drove about 30 miles to the west side, until we found an old rocky dirt road that hunters used to get to the top. As we made the climb, the air was getting real thin, so we set up a base camp. I decided to stay back with a couple of others in case there were any emergencies so we'd be able to go for help. Everyone spread out, and as I watched them ascend into the vast mountain range, I once again understood the true meaning of looking for a needle in a haystack. It didn't matter, though, because they were all so upbeat that I just knew if it was there, this magnificent group of people was going to find it. We sat and watched them for a short time until I just couldn't take it any longer. Who was I trying to kid? I had to do something to help with the search, so we jumped into the four-wheel-drive Jeep and took off down a different dirt road.

The booster from Space Shot 2004.

Space Shot 2004 close-up in rocks.

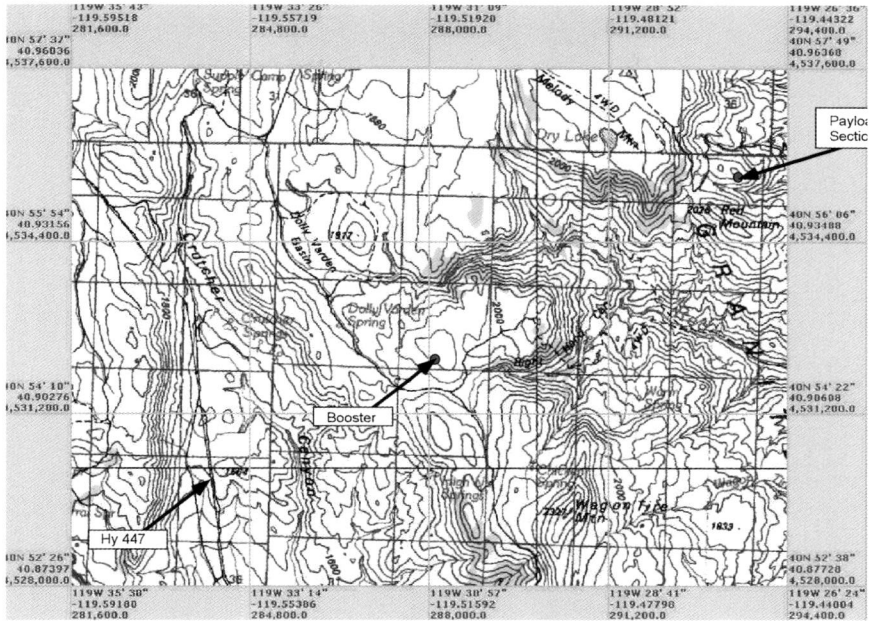

This is where the rocket booster and payload section were recovered from Space Shot 2004.

Just as we were coming up the side of the mountain and around the bend, we spotted a truck that belonged to several members of the search party. It was parked down in a valley, so we headed toward it. I knew it had a ham radio, and I wanted to check on their progress. I was just about to push the microphone button when Ian's voice blasted through, and he screamed excitedly, "We found it, we found it! It's about 6,000 feet up on the side of the mountain. Looks like it came down so hard it actually shattered a couple of rocks into a million pieces and is lodged about four feet into the ground. It's still in really good shape, though, but almost all the paint is burnt off."

Ian examines the heated rocket.

Team members recover the payload section from the side of the mountain.

REND THE SKY

Jerry and Ky carry the recovered Space Shot 2004 payload.

That was incredible news, because that proved the heat it generated was from traveling at supersonic speeds. We had everything we needed to prove we'd made history, and it was just an unbelievable moment. I shouted back into the mike, "This is awesome! Excellent job, guys, excellent job. See if you can get it loaded up, and we'll meet you back at Bruno's!"

Jodi and I took a chair, and I tried to calm myself while we waited for them to get there. When they walked through that door, carrying our payload, my heart went into overdrive. It was almost as exciting as it had been when I pushed the launch button. We shoved some tables together, and they set it down for everyone to see. There were handshakes and pats on the back, and in jubilation, I said, "We need to see what we've got here." I was anxious to download the information from the three flight recorders, but in order to do so, first I had to disconnect the payload section from the electronics section. As I peered inside, I could clearly see how the heat had melted all the plastic that had touched the inside of the nose cone. I could also see a ton of space memorabilia crammed inside, so I asked Jerry, "What the heck is all this junk in here, man?"

Stuff we launched into space.

223

He laughed and said, "Well? That's what happens when you guys leave me behind. I wasn't sure what you wanted to send to space, so I just shoved everything you left on the table inside."

I stuck my hand in and just kept pulling more and more stuff out. There were rocks, carabineers, calling cards, pictures of the Michaelson motorcycle, flags, mail, three cans of Go Fast sports drink, and all kinds of other odds and ends. Everyone just cracked up as the pile got bigger and bigger. Thankfully, we'd all have a little something to take home with us to commemorate the momentous occasion.

Jerry fired up the laptop and got busy downloading the data. We all stood in awe as the magical numbers came across the monitor, proving the rocket had gone well into space and reached a speed of over 3,420 miles per hour.

That moment in time will forever remain etched in my heart, my soul, and my mind. Our small group of people, whom I am honored to call my friends, had just done something that perhaps no other citizens would be allowed by their governments to do. You could travel the world over and not stumble across a prouder group of Americans than each of us in that room on that day. United we stood, the first amateur rocket enthusiasts in the world to build and successfully fly a rocket into space. It wasn't NASA's Goddard Space Flight Center, it wasn't Wernher von Braun, it was just a bunch of guys and girls, the CSXT team, who just never gave up. To me, our team represents the true American spirit. Our day of glory had finally come, when everything happened the way it should have. Before that, everything had happened the way it shouldn't have, and why, we'll never know. But, I kept my faith, we *all* did, and I was blessed to have been able to fulfill my lifelong dream. The only thing that could have made the experience any more meaningful to me would have been having my dear mother by my side. She inspired me more than any other person I have known, or ever will know, in my lifetime. I know she now rests in peace with no more pain, but in my heart I also know she was with me that day. She was there, looking down on me from Heaven, the best seat in the house, saying, "Ky, I knew you could do it. Ky, I knew you would do it." I know that to be true.

Epilogue

By the Numbers

The Civilian Space eXploration Team's Go Fast rocket reached an official altitude of 72 miles, making it the first civilian and amateur rocket to exceed the 62-mile (100-kilometer) international definition of space. Launched on May 17, 2004, at the Black Rock Desert in Nevada at 11:12 a.m., the Go Fast rocket officially entered space at 11:13:41 a.m. PDT, 101 seconds into flight. The maximum altitude was determined by flight reconstruction from data measurements stored on two redundant onboard flight recorders. The official altitude of 72 miles was derived from a high-precision three-axis accelerometer (Crossbow, CXL25LP3) and three-axis magnetometer (Crossbow, CXM113).

Two back-up accelerometers provided additional sources confirming the vehicle exceeded 62 miles. At liftoff, the motor produced 16,000 pounds of thrust and cleared the launch tower accelerating at over 23 g's. After tower clear and a predicted wind-induced yaw maneuver, the vehicle flight performance was nominal and on a nearly straight trajectory as recorded by the onboard magnetometer. At 10.5 seconds into the flight, and with the motor still burning and producing 1,500 pounds of thrust, the vehicle hit a top speed of 3,420 miles per hour and just over mach 5, setting a new amateur speed record. Motor burnout was nominal at 13.4 seconds at an altitude of 49,000 feet with the rocket plume exhaust still visible by ground observers. The canted fins produced a burnout spin rate of eight revolutions per second and put the rocket on a straight trajectory toward space. At 158 seconds from liftoff, the rocket reached a maximum altitude of 379,900 feet (115.8 kilometers or 72 miles) and began its weightless decent back toward Earth. At 240 seconds and still over 250,000 feet in altitude, the primary onboard computer sent a signal to the separation system that immediately separated the payload with the flight recorders from the booster section. Both parachutes, one on the booster and the other on the payload section, deployed nominally.

Decelerating at a peak of 5.5 g's at 160,000 feet, both objects became subsonic at approximately 110,000 feet, where both parachutes were fully deployed. At 850 seconds, the payload section impacted on a mountainside at 62 miles per hour nose-first, 20 miles from the launch site. During the booster's descent, at approximately 50,000 feet, the recovery system malfunctioned for unknown reasons, sending the 211-pound inert booster nose-first, impacting the ground at a terminal velocity of 511 miles per hour.

The FAA's Office of Space Transportation AST-200 Licensing and Safety Division conducted an extensive analysis of the recovered flight data and provided CSXT with this statement on February 28, 2005:

"On Monday, 17 May 2004, the Civilian Space eXploration Team (CSXT) launched the Go Fast rocket at 11:12 a.m. PDT from the Black Rock Desert. The Go Fast rocket was an amateur rocket and therefore did not require a license to launch. However, a waiver to enter national airspace was required and was granted. Two FAA/AST safety inspectors were present for this launch to ensure that conditions of the waiver were met. While AST did not independently verify the maximum altitude attained by the rocket (as tracking radar was not present), post-flight analysis based on CSXT's accelerometer data concurs with CSXT's stated maximum altitude of roughly 72 miles. AST's post-flight results were generally within 5 percent of those of CSXT. Public safety was maintained as both sections of the rocket impacted harmlessly, roughly 20 miles from the launch site."

The Civilian Space eXploration Team would like to thank the office of AST-200 for its significant contributions in assisting public safety assurance and facilitation of approvals for this flight and previous space launch attempts spanning nearly seven years of cooperative effort. We would also like to thank the many sponsors and supporters that have assisted in this endeavor over the years, including Fuscient LLC and Go Fast Sports, Inc., for their unwavering support throughout the Go Fast rocket development, launch, and recovery.

Appendix A

Rocket-Powered Vehicles Built by Ky Michaelson

Ky Michaelson designed and built the *Bonneville Boss* truck. The land speed car that adorns the top of the hauler was a fiberglass model. The real car, named the *Proud American*, was never built.

Pollution Packer dragster. I modified the rocket chamber out of the *X-1* Rislone rocket car that was designed by J. C. McCormick. I made an all-new fuel and pressure system, larger tank, increased the fuel flow, etc., and increased the thrust from 2,500 pounds to 3,000 pounds, and then installed the propulsion system in my top gas dragster. The car was sponsored by the Tony Team.

Jim Hodges' Gator Man *Alabama Express* Funny Car with a 3,500-pound-thrust motor. The car was sold to Fred Goeske somewhere around 1980.

Pollution Packer. I worked with Dick Keller to build a new 3,500-pound-thrust motor for the *Pollution Packer* and later sold this car to Tony Fox. The car was destroyed in a crash.

Pollution Packer Bonneville Proud American dragster monocoque car. I worked with Dick Keller and Tim Kollack for about a year on this project, until I left the Tony Team to venture out on my own. Shortly afterward, the car was shipped to RB Automotive in Kenosha, Wisconsin, where it was finished. The car was driven by Dave Anderson, Vern Anderson, Sammy Miller, and Gerard Brennan. It was later sold to Gary Brammer and Joe Williams, and in 2004, it was sold back to me. It is now parked in my shop in Bloomington, Minnesota.

Ky Michaelson *Miss STP Rocket Dragster*, driven by Paula Murphy. I used the rocket chamber out of the *Pollution Packer*, changed the catalyst pack, made an all-new high-flow fuel system, and increased the thrust to 3,275 pounds. The car was destroyed in a crash, and I later sold the *X-1* Rocket motor to Fred Goeske.

Rocket backpack for Captain Roller Ball. A 300-pound-thrust hydrogen peroxide motor, and I still have it.

Jerry Hehn's *American Dream* rocket car. A 3,500-pound-thrust hydrogen peroxide motor. The car was crashed and then later sold to Fred Goeske.

Super Joe Einhorn's *Space Cycle*. A 1,200-pound-thrust hydrogen peroxide motor. I still have it.

Lee Smith's rocket-powered Arctic Cat snowmobile. A 1,500-pound-thrust hydrogen peroxide motor. I still have it.

Lee Smith's rocket-powered Arctic Cat Kitty Cat snowmobile. A 250-pound-thrust hydrogen peroxide motor. I still have it.

George Lavigne's *Thunderbolt* rocket-powered go kart. A 1,500-pound-thrust hydrogen peroxide motor. The kart was destroyed in a crash.

Pat Best's rocket-powered go kart. A 1,500-pound-thrust hydrogen peroxide motor. The kart was sold to Rosco McGlashan; the *Aussie Invader* is in an Australian museum.

Ky Michaelson rocket-powered hang glider. A 250-pound-thrust hydrogen peroxide motor. I still have it.

Lee Smith's rocket backpack. A 300-pound-thrust hydrogen peroxide motor. I still have it.

Doug Brown's sports car. A 1,500-pound-thrust hydrogen peroxide motor. The car was sold to Dick Williams.

Fred Goeske's Funny Car. A 3,500-pound-thrust hydrogen peroxide motor. I don't know what happened to the car.

Allen Hudson's *Texas Starship* dragster. A 5,000-pound-thrust hydrogen peroxide motor. The car was originally sold to Fred Goeske, then later sold to Rod Phelps, and is currently in the Don Garlits Museum of Drag Racing.

ROCKET-POWERED VEHICLES BUILT BY KY MICHAELSON

A 300-pound-thrust hydrogen peroxide rocket-powered outboard boat motor. I have it.

Fred Goeske's 250-pound-thrust hydrogen peroxide retro rocket reverse motor. Fred has it.

Kitty O'Neil Mattel rocket dragster. A 5,000-pound-thrust hydrogen peroxide motor. I have the car.

Kitty O'Neil Mattel rocket drag boat. A 5,000-pound-thrust hydrogen peroxide motor. The entire boat was stolen.

Human Fly rocket cycle powered by two 1,500-pound-thrust hydrogen peroxide motors. It is in my shop in Bloomington, Minnesota.

Lee Taylor's rocket boat. I designed the 5,000-pound-thrust hydrogen peroxide motor. The boat was destroyed in a crash.

Kitty O'Neil *Rocket Kat* rocket-powered 1977 Corvette. I built the 5,000-pound-thrust hydrogen peroxide motor. The car was destroyed in a crash.

Ky Michaelson/Kitty O'Neil land speed record car. The car was used in Kitty O'Neil's life story, *Silent Victory*, but was then stolen from a garage in Fillmore, California.

Frank Huzar built a rocket-powered 1978 Corvette for Germain Nitchman. He bought the *X-1* rocket motor from Fred Goeske, and then I built a new propulsion system and installed it in the car. The car was sold to Brent Fanning, who changed the body to a Corvette roadster and renamed it *Outer Limits*. From there, the body was later changed again and the car renamed *Concept 1*. At that point, it was driven by Larry Bostic and ran the exhibition circuit from 1979 to 1984. I now have this car.

Dave Henderson and Ed Ballinger's *XL-14 Galactic Crusader* rocket dragster with a 5,000-pound-thrust hydrogen peroxide motor. Fred Goeske now has the car.

Dave Henderson and Ed Ballinger's *Astron Invader* rocket dragster, which was a Kitty O'Neil car with a 7,500-pound-thrust motor. I now have this car.

Fred Goeske's Plymouth Arrow pickup truck with a 3,500-pound-thrust hydrogen peroxide motor. Fred has this truck.

Rocket-powered wheelchair with a 250-pound-thrust motor, parked and ready for me in my shop.

Rocket-powered go kart with a 500-pound-thrust nitrous-oxide hybrid motor, parked in my shop.

Rocket-powered sled with a 250-pound-thrust motor, also parked in my shop, just waiting for the next big Super Bowl party.

Rocket-powered luge with a 750-pound-thrust motor, parked in my shop.

Rocket-powered lie-down skateboard, with 20 staged motors, at my home in Bloomington.

ROCKETMAN

S.S. Flusher, America's secret weapon and one-of-a-kind rocket-powered toilet, powered by a 100-pound-thrust hybrid rocket motor, ready and waiting in my shop.

Rocket bike powered by a 250-pound-thrust hybrid rocket.

Rocket-powered trike powered by a 500-pound-thrust hybrid rocket and hand-built by my young son, Buddy, and me, destined to become a classic Michaelson collector's item if Buddy has anything to say about it.

Hybrid rocket-powered scooter.

1964 Twin T16 Turbonique rocket-powered motorcycle. I sold the motors to Jeff Christianson.

Rocket belt powered by a 500-pound-thrust hydrogen peroxide rocket. I have it.

Rocket bike called *Gizmo* powered by two 250-pound-thrust hybrid rocket motors. I still have it.

Rocket chair powered by a 500-pound-thrust hydrogen peroxide rocket. I have it.

Backpack helicopter powered by two 50-pound-thrust hydrogen perxoide rockets. I have it.

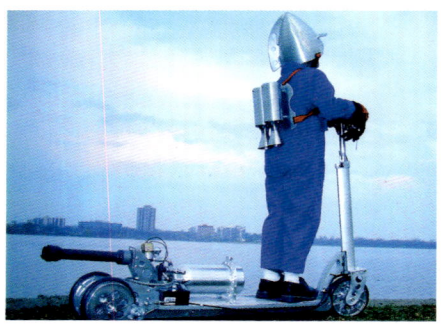

Buddy Michaelson shows off his hybrid rocket-powered scooter.

Ky's wife, Jodi, models the Rocketman rocket belt.

Appendix B

How Hydrogen Peroxide Rocket Engines Work

Now we get down to the nitty gritty, and I take you into the behind-the-scenes innermost detail on how these things really work.

Catalyst Screen Treatment for 90 Percent Hydrogen Peroxide Rocket Motors

Treat the (10), (40), and (60) mesh screens as follows:

A. Apply silver plating to the nickel wire screens with a rough finish to a thickness of 0.001 to 0.002 inch. The plating process must be one that will survive the heat generated in the motor. Plating processes developed for constructing fuel cells will work. This involves application of an intermediate coating (copper) to improve the silver-nickel bond.

B. The plated screens then have to be treated as follows:

Equipment Required:
- A clean oven, set to 850–900 degrees Fahrenheit.
- Stainless-steel racks or trays to hold screens in oven.
- Stainless-steel trays to hold solutions.
- A 10 percent by weight samarium nitrate solution.
- Nitric acid, about 10 percent concentration.

1. Dip screens in nitric acid to remove organic impurities.
2. After cleaning screens, handle only with clean gloves or tongs.
3. Rinse thoroughly in distilled water.
4. Heat screens to 850–900 degrees in oven.
5. Immerse screens in samarium nitrate solution while still hot.
6. Return to oven, heat for 30 minutes after oven reaches 850–900 degrees.

7. Repeat steps 5 and 6 a minimum of four times.
8. A dull straw color indicates coating is applied properly and should be uniform over the screens.

C. Keep the screens stored in a clean container until ready to install in the motor or gas generator. Handle with gloves; do not allow foreign matter to drop into the pack during assembly. Take care so that pipe sealant, Teflon tape, lubricants, etc., do not pass downstream into the catalyst pack.

The Propulsion System

The propulsion system is composed of three basic subsystems. The fuel is 90 percent hydrogen peroxide. No other chemical takes part in the thrust-producing process. This fact puts the rocket motor in the liquid-monopropellant classification of rocket systems. The hydrogen peroxide is a clear, odorless, nonvolatile, nonexplosive, nonflammable, and nontoxic liquid. Hydrogen peroxide is easily decomposed by most materials into water and oxygen gas. This decomposition produces energy in the form of heat, which is what's utilized in generating the power in a rocket motor. Because the fuel is easily decomposed, all materials it comes in contact with must be of the type with which it does not react. The fuel-system materials are 300-series stainless steel, Teflon, Tygon, and polypropylene. These materials must be kept free of such contaminants as dirt, oil, grease, or other organic materials.

Hydrogen peroxide is purchased in 26-gallon drums made of a special alloy of aluminum and is not removed from these drums until time for use. It is then transferred from storage drums to the fuel tank by a polypropylene pump. In some cases, when a specific volume of fuel is required for specific vehicle performance characteristics, this predetermined volume is measured out in a separate polypropylene bottle and then transferred to the fuel tank. Care must be taken not to spill any of the fuel on the vehicle, since it is very corrosive.

The fuel tank is fabricated out of 316 stainless steel. It is shatterproof and rated at 800 psi, with all interior welds sandblasted and cleaned with nitric acid and rinsed with distilled water. The fuel is transferred from the pump to the fuel tank in Tygon hose, and the fuel tank has a pressure vent and a rapid-drain system.

The control system is composed of storage tanks for nitrogen gas, plumbing for distributing the gas, a nitrogen pressure regulator, burst diaphragm, and the throttle valve. The nitrogen gas is used to push the fuel from the fuel tank into the rocket motor. Nitrogen is an inert gas, which is colorless, odorless, nontoxic, and nonflammable. It is stored at a pressure of approximately 2,500 psi in a high-pressure tank

HOW HYDROGEN PEROXIDE ROCKET ENGINES WORK

that has about 2 cubic feet capacity. The nitrogen high-pressure tank is filled from standard-type gas bottles used in welding. The nitrogen storage containers are rated at 3,000 psi, as is all plumbing used in the distribution system. You can also use a scuba tank compressor that produces filtered breathing air in lieu of the nitrogen option—although scuba tank compressors are an expensive investment, they are more convenient than having to carry heavy nitrogen bottles and are much less expensive in the long run.

When the rocket motor is to be fired, the high-pressure nitrogen gas is released from the high-pressure nitrogen tank by a hand-actuated ball valve and allowed to proceed to the nitrogen regulator, where the pressure is reduced and maintained between 425 psi and 650 psi on the downstream side of the regulator. This pressure regulator is a Victor model GD-97, which is used in many space-age applications. The downstream pressure loading is set by the driver while in the vehicle just before the run and can be varied by means of a valve. By actuating another valve, the driver can deliver the nitrogen gas to the fuel tank and pressurize it to the required pressure of 425 psi. Although 675 psi is the optimum fuel pressure at which to operate, the driver may not wish to load the fuel tank to the maximum of 675 psi; if the pressure goes above 675 psi, the safety burst disc on the tank will burst, releasing all the tank pressure. This safety device prevents overloading the fuel tank.

Controlling the flow rate of the fuel into the motor regulates the thrust of the motor. This control is accomplished very simply by a Hills McCanna ball valve, which is located in the fuel line between the fuel tank and the rocket motor. As an added safety device, there is a secondary air-actuated valve in front of the Hills McCanna ball valve that can be used in case the main throttle ball valve should stick open. It can also be used if the front end of the car should become airborne, because it would automatically shut the fuel flow to the rocket motor off, preventing the car from flying.

The main ball valve is located in the fuel line between the fuel tank and the rocket motor. The valve handle is connected to the throttle in the driver's compartment by means of a series of linkages. The foot throttle actuates the ball valve, and the more you open it, the more fuel flows into the rocket motor. If the valve is wide open and the fuel pressure is set at 675 psi, the rocket motor will operate at full thrust. The fuel flow takes place when the valve is opened because the fuel tank has been preloaded to a given pressure, but the nitrogen gas and this loading are kept constant during the run by the nitrogen regulator. When the valve is closed, the fuel flow to the motor is stopped and the motor shuts down instantly. After firing the motor, the pressure in the fuel tank is released by means of a vent valve actuated by the driver. This will also stop the rocket motor from running in an emergency situation.

The rocket motor is constructed of nonmagnetic 316 stainless steel, and the catalyst pack is a mix of silver wire screens packed into a mat and positioned just behind the fuel injector plate. The principle of the operation is simple. The fuel enters the fuel manifold and is evenly distributed over the catalyst pack by a plate perforated with hundreds of small holes, much the same way water comes out of a showerhead. This spray of fuel is forced through the maze of the catalyst pack and is completely decomposed to superheated steam and oxygen gas. This decomposition liberates a great deal of heat, causing the temperature inside the motor to reach 1,370 degrees Fahrenheit. In turn, the temperature rise causes the pressure inside the motor to soar to 300 psi, and the combination of heat pressure results in the steam and oxygen being forced out the exhaust nozzle at a speed of 4,000 feet per second. At full thrust, a 2,500-pound-thrust rocket motor will consume 3 gallons of fuel per second, with each gallon weighing 11.6 pounds. Since the decomposition of the 34.8 pounds of gases per second also produces 34.8 pounds of gases per second, a situation exists where the motor is expelling 34.8 pounds of gas at the enormous speed of 4,000 feet per second. This results in a forward shove. For every action, there is an equal reaction. If the throttle valve is only partly opened, the driver only gets partial thrust, thereby making it possible to drive the vehicle at any speed desired up to the maximum.

The effect of fuel decomposition by the catalyst pack is very similar to throwing a bucket of water on a red-hot chunk of metal. The water instantly flashes to steam, as does the peroxide when it is forced through the silver and nickel of the catalyst pack. The exhaust from the rocket motor is not visible and, since it is composed of steam and oxygen, is also not toxic. Because there is no burning taking place in the motor, there is no flame produced. The only sign of action on the part of the rocket motor is a great deal of noise.

Hydrogen Peroxide Rocket Schematics

Hydrogen Peroxide Rocket Motor System Used in Ky Michaelson's Rocket Car
1. 680-psi relief valve
2. H_2O_2 dump valve, must be on bottom of tank
3. Rocket motor
4. H_2O_2 thrust-control valve, manual
5. Relief check valve
6. H_2O_2 shut-off valve pneumatically operated and electrically connected to the front wheels and parachute
7. H_2O_2 tank assembly, 316 stainless steel, 800-psi working pressure

HOW HYDROGEN PEROXIDE ROCKET ENGINES WORK

8. H_2O_2 tank vent valve (shut-off valve), manual vent
9. H_2O_2 tank pressure gauge
10. H_2O_2 burst disc assembly
11. GN2 pressure check valve
12. H_2O_2 fill valve on top of 7
13. Dome pressure gauge, 0–1,000 psi
14. Dome regulator, manual pressure and bleed w/dome gauge
15. GN2 regulator, manually or dome loaded
16. High-pressure manual isolation valve
17. GN2 fill and dump valve
18. GN2 pressure gauge assembly to 3,000 psi
19. Burst disc assembly, burst pressure 2,200 psi
20. GN2 tank (two required, 12-inch fiberglass-wrapped stainless steel with pressure of 3,000 psi. Proof tested at 5,000 psi with 2,200 psi reduction valve)

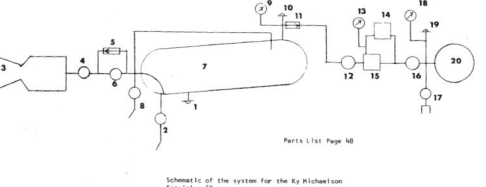

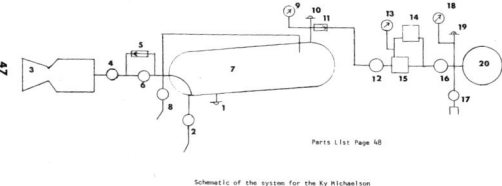

Schematic of the system for the Ky Michaelson Special. 60

Parts List for Hydrogen Peroxide Rocket Car Motor

A. N2 filler line vent valve
B. N2 filler line valve
C. N2 filler line pressure gauge
D. N2 filler line connector
E. N2 fill connector
F. N2 pressure gauge
G. Burst disc 3,000 psi
H. N2 shut-off valve
I. Dome loader valve
J. N2 pressure gauge
K. Dome vent valve
L. Relief valve 650 psi
M. N2 regulator
N. Check valve
O. Relief valve 650 psi
P. Dome vent valve
Q. Pressure gauge
R. Burst disc 1,000 psi
S. H_2O_2 fill
T. H_2O_2 drain plug
U. Check valve
V. Safety shut valve
W. Battery
X. Safety system switch
Y. N2 pressure gauge
Z. Solenoid valve
A-1. Relief valve 2,500 psi
B-1. H_2O_2 tank

Appendix C

Official NHRA Rocket Car Rules and Regulations

Now that you've read up on all those early legendary rocket racers, I decided I'd better be sure to provide a better understanding of exactly what we all went through to obtain permission to compete on the NHRA tracks. The rules below were very strictly enforced and adhered to, and once you read them, you'll have a much better understanding of what a race team and the vehicle it supported really went through to thrill so many people. It was far from easy!

National Hot Rod Association Procedure and Operational Acceptance Requirements, Rocket-Powered Exhibition Vehicles

Introduction
Outlined here are the minimum operational and safety specifications and standards that rocket-powered vehicles must meet before being considered for exhibition operation on NHRA-sanctioned tracks.

General Requirements
A. It shall be the full responsibility of the owner of the vehicle to assure that only qualified personnel who have been fully checked out with the specific vehicle and are directly associated with the car be allowed to fuel, check out, and operate the vehicle. Further, it is the responsibility of the owner/builder to have met all safety specifications as outlined by the NHRA technical committee.
B. Approval and certification of each and/or any rocket-powered exhibition vehicle must be done on an individual basis through the special Exhibition Technical Committee of the National Hot Rod Association.
C. The owner(s) of each rocket-powered vehicle must submit in writing, together with necessary schematics and data, a request for NHRA approval and license to operate on NHRA-sanctioned drag strips.

OFFICIAL NHRA ROCKET CAR RULES AND REGULATIONS

D. The rocket-powered vehicle will then be submitted for inspection, along with all necessary certificates, to the special NHRA Committee for approval. This inspection will take place at the Pomona Drag Strip in Pomona, California. A $300 license and inspection fee will be charged and collected at the time of inspection.

E. Reliability test runs and, if necessary, driver license test runs, will be scheduled and arranged for by the owner(s) of the rocket-powered vehicle at their expense at a suitable approved track or facility. No spectators or audience shall be in attendance at these firings. Arrangements shall also be made with NHRA for a qualified observer to be present during these test runs.

F. If all is in order, the rocket-powered vehicle will be sealed, license papers awarded, and approval given. At that point, bulletins and all necessary announcements would be made concerning the vehicle.

G. Any violation of safety regulations and requirements or careless operation of the vehicle may be cause for suspension of said vehicle and/or driver from operating on NHRA-sanctioned tracks.

Chassis and Running Gear

A. Chassis and cage area must meet or exceed all current dragster standards as outlined by NHRA and SEMA. Due to the extreme speed potential and the aerodynamics involved, no Funny Car bodies will be allowed at this time.

B. Total vehicle weight, excluding fuel and driver, must not exceed 1,200 pounds.

C. Driver compartment must be sealed off from the rest of the car; fluid tight. Lines passing by driver must be enclosed in armored conduit and sealed; fluid tight. Hoses, lines, and valves shall not be allowed in driver compartment.

D. Wheels and tires shall be tested to twice the recommended maximum tire pressure. In no case shall tire manufacturer data limits be exceeded in speed or loading.

E. Dual disc brakes on both rear wheels are minimum requirement.

Engine, Tanks, and Fuel System

A. Must be of a design to utilize monopropellant (liquid) fuel only.

B. Pressurization system must be built according to the schematics, with design details illustrating safety factors utilized and system fail-safe provisions.

C. Pressure sphere(s) must be proof-tested at two times (2x) maximum working pressures. NO YIELD.

D. Schematics and detailed designs of the propellant-fueled system must be provided showing safety factors utilized and system fail-safe provisions. Proof-test propellant tanks at two times (2x) maximum working pressure. NO YIELD.

E. Schematics and detailed designs of the rocket engine system must be provided showing the design safety factors utilized and system fail-safe provisions with the nitrogen-purging and automatic-shutdown capabilities. Wall thickness of motor shall have a yield factor of three times (3x) working pressures.
F. All components shall be fabricated from low-carbon, nonmagnetic, easily weldable stainless steel; i.e., for H_2O_2 systems, all materials shall be either 304, 316, 321, or 347 stainless steel. NO dissimilar materials shall be used in the system that will support electrolytic corrosion. NO material shall be considered that is subject to shattering or stress corrosion. All welding is to be done by a certified welder (certification required). All pressure vessels must be hydrostatically tested and certified to required specifications.
G. Components, lines, etc., downstream of the regulator shall be cleaned and passivated prior to assembly. The regulator shall be fabricated from compatible materials with a propellant liquid and vapors and shall be LOX clean prior to assembly in the system. The pressurization system upstream of the regulator must be LOX cleaned prior to assembly of the system. The pressurization system does not necessarily have to be compatible with the vapors and/or liquids, provided the operational procedures guarantee that the pressurization system is always provided with a higher pressure than the propellant feed system; i.e., always provide a positive pressure in pressurization system. NO REVERSE FLOW.
H. A blow-out diagram or pressure disc will be provided on the propellant and pressurization tanks to guarantee that the working pressures cannot be exceeded.
I. The safety valve and/or burst disc on the propellant tank must be routed through tubing or ducting in such a manner as to direct gases and/or propellant rearward. It is important to note that the safety valve or burst disc be located on the ollage side (pressure side) of the propellant tank; never on the H_2O_2 side.
J. The propellant tank (H_2O_2) must be equipped with a pressure gauge. The gauge on the regulator is not sufficient.
K. Concentrated H_2O_2 (hydrogen peroxide) rocket-powered vehicles must provide a normally open vent on the propellant tank at all times, except when pressurized.
L. The system and components must not trap H_2O_2 in any of the hardware or instrumentation.
M. A system must be provided that allows complete draining of the system after the operation of the vehicle. Propellant must never be left in the system or tanks, or returned back to the transport containers after operation of the vehicle.
N. Only oil-free gaseous nitrogen shall be used to pressurize the system, using only LOX clean connectors and/or flex lines with no substitutions.

OFFICIAL NHRA ROCKET CAR RULES AND REGULATIONS

Fuel

At this point and until further experience with thrust-propelled vehicles is gained, the National Hot Rod Association will limit propellant to hydrogen peroxide (maximum 90 percent commercial H_2O_2). No pre-heating of fuel (H_2O_2) is allowed. Hot drums of fuel must be disposed of immediately.

Driver and Crew
A. Must be licensed as a unit with the vehicle to be driven.
B. Must make initial test runs at a suitable approved track or facility with no spectators or audience in attendance.
C. Standard NHRA license procedures will be approved. Six runs in two separate sessions, blindfold test, and FAA type II physical examination required.
D. Each car must have at least one (1) trained crew chief in addition to the driver when operating the car. Crew chief must be thoroughly checked out on operational procedures.
E. All crew members must have proper apparel as listed below:
   1. Polyethylene jumpsuit.
   2. Rubber or polyethylene boots.
   3. Rubber or polyethylene gloves.
   4. Clean Dacron underclothing.

Safety Regulations
A. A minimum of two braking parachutes is required. The primary chute must be on an integral control circuit with safety valves and a chute control line attached to driver's wrist. The secondary chute will be used as an independent backup system.
B. When the car is fueled and being pushed to the starting area, all effort must be made to keep the area behind the car (up to 50 feet) clear of all persons.
C. A 50-foot area must be roped off for the rocket car pit area. Only authorized persons should be allowed in the area.
D. When fueling the car, it is mandatory that a water supply of at least 50 gallons be available (hose, fire truck, or large 55-gallon drum with portable pump).
E. An airborne safety device shall be incorporated into the front suspension system that will automatically vent the propellant and pressure tanks, close the secondary throttle valve, and close the valve between the high and low pressure tanks in the event the front wheels terminate ground contact.

F. Throttle valves must be manually operated and must be constructed in such a manner that they are normally closed, and the vent valves must be normally open to meet fail-safe conditions.
G. Test firings of the rocket-powered vehicle will be accepted in lieu of laboratory passivation techniques. Three (3) firings will be required starting with 35 percent H_2O_2 and ending with 90 percent. These reliability tests must be observed by an authorized official and conducted at a track or acceptable facility with no spectators or audience in attendance.

Limits and Suspensions
A. An absolute speed limit of 330 miles per hour must be observed. Violation of this limit will result in a one (1) year suspension of the unit (car and driver) from operation on NHRA tracks.
B. No side-by-side exhibition or competition runs will be allowed at this time. A one-year (1) suspension of the unit (car and driver) will be the result of violation.
C. Any car or crew using any fuel more potent than 90 percent H_2O_2 will be suspended from operating on NHRA tracks indefinitely. This includes the use of hybrid plastics or any other power-boosting fuels.

Note: The above suspensions are automatic. Appeal may be made through the National Hot Rod Association at the expense of the violator. A contest board consisting of at least three car owners and/or drivers and a representative of NHRA would hear the appeals.
Revised: January 23, 1974